Health & Hazards
in a
Changing Oil Scene

PROCEEDINGS OF
THE INSTITUTE OF PETROLEUM
LONDON

HEALTH & HAZARDS IN A CHANGING OIL SCENE

Proceedings of
The Institute of Petroleum
1982 Annual Conference
London, UK

Published on behalf of
The Institute of Petroleum, London

JOHN WILEY & SONS
Chichester · New York · Brisbane · Toronto · Singapore

British Library Cataloguing in Publication Data:

Institute of Petroleum. *Annual Conference* (1982: London)
   Health and hazards in a changing oil scene.
   1. Petroleum products—Safety measures—Congresses
   I. Title
   363.1'89   RA578.P/

   ISBN 0 471 26270 6

Printed in Great Britain by
St Edmundsbury Press, Bury St Edmunds, Suffolk

CONTENTS

ORGANIZING COMMITTEE

Chairman	C.N. Thompson, CBE, Shell UK Limited
Members	P. Jones, Institute of Petroleum
	W.L.B. Leese, Britoil plc
	G.R. Mayhew, Institute of Petroleum
	A.L. Mills, Burmah-Castrol Company
Conference Secretary	I.A. McCann, Institute of Petroleum
Publications Secretary	C.H. Little, Institute of Petroleum

LIST OF CONTRIBUTORS

ALDERSON, Dr. M.R., Office of Population Censuses and Surveys, St. Catherine's House, 10 Kingsway, London, WC2.

BARRELL, A.C., Health and Safety Executive, 25 Chapel Street, London, NW1 5DT.

BLACK, Sir Douglas, Royal College of Physicians, 14 St. Andrew Place, Regent's Part, London, NW1.

DAVIES, Dr. D.M., British Petroleum Company Limited, BP Group Occupational Health Centre, Chertsey Road, Sunbury-on-Thames, Middlesex, TW16 7LN.

DAVIS, Prof. P.R., Department Human Biology and Health, University of Surrey, Guildford, Surrey, GU2 5XH.

DOLL, Sir Richard, Regius Professor of Medicine, Green College, The Radcliffe Observatory, Oxford, OX2 6HG.

DUNCAN, Dr. K.P., Director of Medical Services, Health and Safety Executive, 25 Chapel Street, London NW1 5DT.

EDDERSHAW, B.W., ICI plc, Petrochemicals and Plastics Division, Wilton Works, Middlesbrough, Cleveland, TS6 8JB.

ELLIOTT, Dr. D.H., Shell UK Limited, UKPMO/3, Shell Mex House, Strand, London, WC2R ODX.

EYRE, Dr. J.A., Shell Research Limited, Thornton Research Centre, P.O. Box 1, Chester, CH1 3SH.

FULLER, H.I., Rookery Farm, Steventon, Abingdon, Oxon.

GOLDEN, Surg. Cmdr. F. St. C., Institute of Naval Medicine, Alverstoke, Gosport, Hants., PO12 2DL.

GREGORY, A.T., BP Oil Limited, Britannic House, Moor Lane, London, EC2Y 9BU.

GRIEVE, A.M., Shell UK Oil, Training Department, Shell Haven Refinery, Stanford le Hope, Essex.

JONES, P., Institute of Petroleum, 61 New Cavendish Street, London, W1M 8AR.

JOYNER, Dr. R., Shell Oil Company, 1 Shell Plaza, P.O. Box 2463, Houston, Texas 77001, U.S.A.

KLUGE, Dr. A., DGMK, Nordkanalstrasse 28, 2000 Hamburg 1, Federal Republic of Germany

LEESE, Dr. W.L.B., Britoil plc 150 St. Vincent Street, Glasgow G2.

NELSON NORMAN, Prof. J., Institute of Environmental and Offshore Medicine, Aberdeen University, 9 Rubislaw Terrace, Aberdeen, AB1 1XE.

RUSHTON, Dr. L., Thames Polytechnic, Wellington Street, London, SE18.

SMITH, R.H., Esso Petroleum Company Limited, Esso Research Centre, Abingdon, Oxon, OX13 6AE.

WEAVER, Dr. N.K., American Petroleum Institute, 2101 L. Street Northwest, Washington DC 20037, U.S.A.

CHAIRMAN'S INTRODUCTION

Sir Douglas, Ladies and Gentlemen,

I am very pleased to open this Institute of Petroleum Conference with a few words of welcome to all present and participating in it. I am particularly glad to be doing so in the very appropriate surroundings of the Royal College of Physicians. The Institute is much indebted to the Royal College and to you, Sir Douglas, its President, for making this possible.

The Conference is, of course, an annual event, although next year, with the World Petroleum Congress itself taking place here in London,* we shall defer our own Conference and resume again in 1984. It is a measure of the variety of the interests of the Institute of Petroleum's activities that Conferences may range from, for example, the Financing of Energy Developments, as last year at Cambridge, through Health and Hazards, this year, to the upstream subjects of Geology and the Marine Environment and Offshore Developments in 1984 at Aberdeen. The Institute covers a great range. But two things characterise all its activities and indeed its founding fathers specified them - sense of humanity and of human relationships; and professional objectivity and technical excellence.

No-where do these characteristics more desirably come together than in issues of health and safety. On a personal note, therefore, I am very glad, at the beginning of my Presidency, to be introducing a Conference so exactly tailored to the role of the Institute.

The last Annual Conference on a similar theme of health and safety was held at Eastbourne in 1976, following the 1974 Health and Safety at Work Act. Six years on it is timely to meet and review developments since then in a number of selected key areas. We shall have, this year, the benefit of Dr. Alderson's reporting on the completion of an extensive epidemiological survey, the genesis of which was at that 1976 Conference, covering both the UK refinery and oil distribution sectors and also as a user sector, London Transport Diesel Maintenance Workers

* *The Barbican, 28th. August - 2nd. September 1983*

1

And we shall have also this year a benefit we did not have in 1976, when the speakers and topics were essentially UK-orientated, in that we are privileged on this occasion to welcome distinguished representatives from the DGMK in Federal Republic of Germany and from the API, the American Petroleum Institute. They will give us an account of current activities in the health research field carried out by their own organisations. The Institute has very close links with both organisations and we are very pleased indeed to welcome them to this gathering.

I welcome them, and all our participants; and now ask you, Sir Douglas, to give us your opening key-note address.

London

A.T. Gregory
President, Institute of Petroleum

OPENING KEYNOTE ADDRESS

Sir Douglas Black

President of the Royal College of Physicians,
London

Mr. Chairman, ladies and gentlemen, this is one of those occasions
which are not in fact too rare in the life of the President of this
college where I am really much more conscious of the weight of my
responsibility this morning than I am confident of my ability to
discharge it.

Let me start with the easy end of it. I second the welcome which
your Chairman and President has just given: welcome to this
College. Like many other corporate bodies, real power in the College
of course, does not lie with the President: it lies with the
Treasurer, and it is really the Treasurer who is your host this
morning and for the next couple of days.

Now the worst thing to do in starting a talk of any kind is to
expatiate on how incompetent you are to do it and I am not going to
do that but I am going to invite your sympathy for a certain
difficulty in which I find myself. When you are required to speak
on something on which you are hardly even a layman you can go for
one of two options. You either get into the history of the subject
or you talk about the ethics. Unfortunately for me but perhaps
fortunately for you, I do not know, I gave a Lucas Lecture to
the Faculty of Occupational Medicine. I think it was the first
Lucas Lecture, and in it I exhausted, and probably more than
exhausted, my knowledge of the history, and perhaps the audience
too. When I was speaking to the Society of Occupational Medicine
I spoke about the ethics of occupational medicine; also, the
Faculty, which has really been my mentor in all these things, are
producing now the second edition of a very useful pamphlet on ethics
as they effect the work of occupational physicians.

Now that is the end of the excuses and the apologies. I will now
try to address myself to the subject. I have been considerably
assisted or perhaps potentially so by a hundred or two pages of
material which have been sent to me. It deals mainly with Dr.
Alderson's work, but since Dr. Alderson is obviously so much more
competent to present his work than I am, I am going to leave him
to do it and not make some jejune comments which might merely
embarrass him or even expose him to the necessity of contradicting

flagrantly which would not be pleasant for either of us.

Now I do have to say something, and the first thing I would like to say is that I am extremely conscious of the breadth of interest which lies in the scope of this Institute and the petroleum industry. I think the way it is presented in the media makes people think naturally primarily of the hazards of diving in the North Sea. That is the glamorous end of it, encompassed by all manner of physical hazards and I am remembering that the title of this conference is Health and Hazards in a Changing Oil Scene. But of course that is only the tip of the iceberg. There is also the huge petrochemical industry which I think has pretty well replaced the coal tar industry as a source of many products including fertilisers and all the kind of thing. So it is a very broad scene from the physical hazards and obvious dangers of work on the oil rigs and beneath the surface of the North Sea to detailed chemical problems which arise in the subsequent processing and handling of petroleum. For those reasons it is particularly appropriate that the Institute is a broad-based multi-disciplinary group. I have only had time this morning just to glance at the list of those attending or hoping to attend but already I have seen that there are engineers, there are managers and others as well, of course, as doctors with whom I am perhaps more familiar. So I do recognise the very broad base and how appropriate it is to an Institute such as the Institute of Petroleum.

I would like to take up as my first theme what I will loosely call the anti-science lobby and why it is something to be resisted. The second thing I would then like to go on to is to say something about hazards, and perhaps more the matter of hazards in general. Let me anticipate a little by a reminiscence. One of the first symposia that the Faculty of Occupational Medicine arranged was originally to be entitled "The Acceptance of Risks" and that the title had to be changed to "The Assessment of Risks" because it was thought to be too tendentious so I think we have to take a rather cool and objective look at hazards and that is what I will try to do, bringing up if I can some evidence in support.

I would like to read a short passage from the Lucas Lecture to which I referred and again this is an open manifestation of only partial confidence - when someone is reduced to quoting from their own work they are in a rather bad way. What I said in the Lucas Lecture was this: "The revolution in the past century in attitudes to the health problems of the worker has brought immeasurable benefits to the whole of society and I see no realism in those who hanker nostalgically for the pre-industrial society which in fact they never knew. Even the privileged few in those days, who no doubt had a comfortable life in the main, were still menaced by risks to health which were not understood and so could not be controlled in any way. For the great mass of people, industrialisation has brought opportunities that were unimagined even by the wealthy of former times."

Now just to elaborate that point, think of what it must have been like to suffer from a serious or even perhaps a minor illness in those bygone days. We all know about Charles II and the tortures to which he was subjected by the combination of illness and his doctors. But I would like to give two other examples of the dangers of illness in those days. For the first of these I will draw on Lord Hervey's memoirs which describe that interesting trio - Lord Hervey himself, King George II and Queen Caroline. George II was not a sensitive man and he was not a man of culture except for the honourable exception of music, of course. He was a patron of Handel. If he found his wife reading a book he tended to storm out of the room in a passion or even throw another book at her. She had a fatal illness, as we all do, of course, and this must have been some form of bowel obstruction. Guessing at the illnesses of the past is not an endeavour in which one ever reaches a great degree of certainty. It is much worst even than trying to diagnose illnesses now. But she must, I think, have had some degree of intestinal obstruction because she became totally constipated, had considerable swelling of the abdomen and even a swollen loop of the bowl which, following some surgery, came to the surface and burst. This was a pretty bad scene and George II, that insensitive man, as I have emphasised, was so moved by it that he said "I will never marry again, I will take mistresses." So far as I know, he kept both of those promises to the letter.

My example is the question why Louis XIV was succeeded not by his son, as might be the normal thing, nor by his grandson, but by his great-grandson. Again, there is a relatively simple, probably too superficial explanation, and that is that the great Fagon, the doctor of the Court, managed to eliminate all the intervening generations in the male line. The reason why Louis XIV himself survived to a good age was because he was the man who had the power and capacity and also perhaps the insight to pay absolutely no attention to what Fagan recommended. That is another example of the hazards of those days to which some of us look back very romantically.

There is a widespread lack of appreciation of the importance of science in our modern life. Only yesterday, in "The Times", there was a letter from Professor Margaret Gowing, who is the historian of atomic energy and of much else, the history of science, and she pointed out that although there was in Oxford a course which was designed to bring together scientists and people on the arts side, there were very few takers and, oddly enough, there were more takers from the science side than there were from the arts side. This reminded me to some extent of an initiative that Brian Flowers took when he was Professor of Physics at Manchester. He set up a school of liberal studies in science in the science faculty and this was, again, much more taken up by potential scientists than it was by people from the arts side. Yet I would have thought that someone with the combined kind of education would have been an ideal civil servant on the Fulton model which has never really been fully implemented, and in time become an excellent Permanent Secretary who was not blind in one eye, which is what I really think is the situation with someone who has had very little, or even no scientific component in their education.

We may group the causes of misunderstanding of science as the four "M's" — two lay fallacies and two fallacies which are held mainly by scientists themselves. Obviously there is a spill-over between the two groups but the two lay fallacies I will describe as the Macropaedia fallacy and the Mistique fallacy and the two scientific ones are the Measurement fallacy and the Methodology fallacy. Again, I am quite conscious of the overlap between measurement and methodology but it makes for a neat balance if you have four "M's", two in each category and not just three.

The Macropaedia fallacy: you know of the new Encyclopedia Britannica, which has a section which includes snippets of all knowledge and I think that is how some lay people thing scientists operate. That there is some huge store of scientific knowledge sealed and delivered for all time and that what scientists learn is how to look it up.

Then there is the Mystique fallacy which is that science is such a mystery that no one can hope to penetrate it unless they become full-time scientists. I think scientists themselves have perhaps contributed slightly to that because we have been rather shy about writing for the popular press and there is a reluctance to undertake *haute vulgarisation* of science which I think has stood in the way of public appreciation of science.

The scientific fallacies, in my view at least, are first, that all science is Measurement. Obviously, measurement is a very important component of science but equally there are descriptive sciences like palaeontology, for example, in which measurement plays some part but is by no means the whole story. Secondly, Methodology — again some scientists think that the way to progress is to find a good method, to define it and then to apply it in all sorts of situations. I think these beliefs do give a distorted view of science and, of course, to some extent I naturally caricatured them. I think that like most people who think about the philosophy of science I have been captivated by the Popperian approach which is that science is really about ideas even more than it is about facts and that you proceed not by verification which, again, I am sure would by the lay concept of science but you proceed by repeated attempts at falsification and the more attempts at falsification are resisted the greater then is the generality and presumably the value of the hypothesis which has been defined in that particular way.

It is now time to say something about hazards. We know, of course, that risks of all kinds are present in the world and, to some extent, they have to be accepted. In relation to any particular occupation you have what might be called specific risks which arise directly out of the nature of the occupation and, in addition to that,

you have the general risks which arise out of the life-style which
may be shared of course by the family, the wives and families of the
people in the industry. In other words, there are highly specific
hazards, whether these are physical or chemical and, in addition,
there are the consequences on life-style of the terms and conditions
of service within the industry. One has, I think, to pay some
attention to each. Again, I would like to read a very short
passage from the Lucan Lecture:
"While safety is a laudable and, indeed a necessary aim, towards
which we must constantly strive, it can never be an absolute.
There are risks in everything we undertake and our selection of
those which we choose to accept is, to say the least, paradoxicial
at times. For example, the vociferous campaign on the risks of
whooping cough vaccination must have been responsible for some
mortality and a great deal of unnecessary morbidity. We should
therefore take some care that our precautionary measures do not
induce total inactivity. I would like to exemplify that point
further and bring in perhaps another point. I am sure that the lack
of logic which I tried to illustrate by the whooping cough example
is also shared by lack of attention to the financial aspects of
the situation. I am going to draw on an interesting lecture
which was given, I think, three or four years ago in this very room
by David Kerr, who is Professor of Medicine in Newcastle-upon-Tyne.
He discussed the relative financial risk that people are willing to
accept in order to save a life. The underlying calculation is that
you can cost a precautionary measure and you can also estimate how
many deaths have resulted from absence of the precautionary measure
over an adequate period of time. That enables you to put a price
per life saved. To give some examples: screening pregnant women
to avoid stillbirth is estimated to cost £50 per life saved and that
particular measure was postponed as being too expensive. The next
one is the introduction of child-proof containers and that is
estimated to save lives at an expense of £1000 per life saved. That
one too was postponed for a considerable period, although it has
now become law, again because initially it was considered to be too
expensive. Then compulsory cabs on farm tractors – the cost of
saving a life there is £100,000 and I do not surprise you by saying
that that one was adopted. Then by various combinations of dialysis
and renal transplantations you can buy twenty years of life for
approximately £200,000 and again that has been adopted in part in
the U.K., even if not to the same extent as in many other countries.
Then getting into the big league, as it were, there was the Ronan
Point disaster, and the various strengthening operations that were
set in motion after that. I do not think they were estimated in
advance but with retrospective costing each potential life saved by
that comes out at £10m which is beginning to be quite a substantial
sum of money. Finally, the improved fire precautions in British
hospitals and universities in the 1970s, and the measures which
were used following the disaster at Melbourne to strengthen girdered
bridges, are examples where the value of a life saved is too high to
calculate, certainly over £10m so you see there is no great logic
in things that are adopted or not adopted and I suppose we could all

think of other examples to show that at least we are not the
absolute slaves of economics.

Looking for some thread behind all this, to me at least it seems that
the element of drama is the thing that makes the spur to action,
and of course one can illustrate that in a much more common
observation in everyday life. We accept the day-in and day-out
deaths on the roads but when there is an air crash or even a train
crash, that immediately makes headline news. Something which
happens dramatically and suddenly gets noticed, and may even lead to
action; whereas when things happen gradually and steadily we just
become acclimatized to them.

Moving on, I would like to draw attention to the fact that there
are two underlying disciplines which are not perhaps immediately
or directly relevant to the problems of the oil industry but are a
sort of substrate for them. These are epidemiology and toxicology
and they are both disciplines which have been pursued with
distinction in this country but which, nevertheless, could well do
with further expansion. I would even make a plea that when industry
finds itself with some funds it should not tackle only the immediate
presenting problems but should also give some attention to the
basic sciences which underpin them.

Toxicology and epidemiology are both disciplines of the most vital
importance, obviously, not only to your industry but to industry
in general and the study of medicine in general. I am not going to
say anything about toxicology other than to welcome the quite
modest steps that have been taken elsewhere. The earlier lack of
interest in toxicology was a worry to me when I had some departmental
responsibility for it but I think things are beginning to happen
in this field.

Epidemiology - I am very conscious that some of the leading
exponents of epidemiology in this country are in this room and I am
going to confess that when I want to annoy epidemiologists (which is
something that I rarely want to do), I point out that what they
show is 'association' and not 'causality'. Sometimes, however, the
association is so repeatable and so convincing as in the case of
smoking and lung cancer that no reasonable person could escape the
evidence.

Attitudes to causality depend on points of view. A tobacco
manufacturer is reluctant to see causality of lung cancer in
cigarettes; on the other hand, the anti-fluoride lobby claim
causality when there is not even an established association between
fluoride and cancer.

There are some particular things that I want to say about
epidemiology. First of all, the difficulty as well as the importance
of defining the population which is at risk. That depends, you see,

to some extent, on accuracy of records and it must, I suppose, be
more difficult in a private sector industry comprising many different
and diverse companies than it is in a unified nationalised industry
where presumably the number of people employed is more easily
collated because they are on some kind of centralised payroll. That
is one way to get at particular populations, apart from of course,
censuses. Another point I would make is that there is very
considerable importance in specifying the type of industry activity
in which people are engaged and even the type of materials they are
handling as well as just giving some general term such as "fitter"
or "cleaner". You have to know in much more detail what happens.
This lesson was brought home to me many years ago when I saw a patient
in the Out Patient Department with wrist drop which is a fairly
characteristic symptom of lead poisoning. His occupation was listed
as "demolition worker", but I found that what he was actually doing
was demolishing railway stations, and in particular taking the paint
off metal girders with a blow-lamp. This paint, of course, from the
Victorian era was full of red lead and he was getting his lead
exposure in that way. You have thus to get at the actual specific
job that someone is doing and not rest at the generalities like
"demolition worker". It was fortunate, of course, that in the case
described the clinical presentation was so strongly suggestive of
lead exposure.

We have been, in the College, having some discussion about death
certification and again this is something of the most cardinal
importance for determining what are the specific death risks
associated with particular occupations. Again, there are
difficulties and snags and some of those we encountered in studying
inequalities in health because the occupation that is recorded may
not be precise in the way that I have just been trying to indicate
in the case of the demolition worker. You may not have sufficient
indication from the rather summary statement of occupation on the
death certificate to get the real picture.

You may also wish to have information as to the general risk as
opposed to the specific ones of occupation. There are, however,
in this respect some difficulties of certification. What determines
lifestyle is the total income coming into a household. There are
situations in which the wife may be the major earner and in those
circumstances the husband's recorded occupation does not give you a
full picture of the sort of contribution which affluence or poverty
may be making to ill-health. Again, there are considerable
difficulties of certification of the actual cause of death. This
arises to some extent from the way in which the doctor filling in
the certificate is instructed to do it. What he is told to put first
is the immediate cause of death and then, as a secondary item the
other factors which may have contributed. What is easy to miss out
is an important long term disease like diabetes. If someone dies of
pneumonia it may not be recorded that they had also had diabetes
for many years unless of course the diabetes was an actual
contributor to the cause of death. I do not know whether
Dr. Alderson might not want to comment on that, because he is now in

the thick of the death certificate business, as it were.

I would like to end by saying something about the Occupational Health Faculty which I have already mentioned from time to time; and also about the role of the College in relation to the problems of occupational medicine. The College has a very general responsibility and our central aim really is to try to ensure high standards of medical practice in the interests of patients. We do this by various ways. We have an examination which George Pickering in his unparalleled way described as "a difficult examination at an elementary level". By that cryptic phrase we mean that it is not something that certifies that someone is fully qualified physician. What it certifies is that they have attained the sort of general grasp of medicine or one of its specialities that indicates that they are fit to go on to higher medical training. We have a number of committees on the different specialities within medicine and also a joint organisation with the Scottish Colleges called the Joint Committee on Higher Medical Training. The College quite early recognised that there were very particular problems about occupational medicine. For one thing it is the largest group of doctors practising without the NHS network and, for another the sort of training conditions and programmes are really very different from those of the average hospital doctor in whatever speciality. So we were very pleased to set up the Faculty. I have made that sound as if the College has displayed tremendous vision. Of course, the College does have tremendous vision but specifically there was vision among Fellows of the College engaged in occupational medicine, some of whom are present today. We are very glad and very proud to have under our auspices at the College the Faculty of Occupational Medicine.

I have started off by talking rather loosely about science and about its application but I think that really it is a false antithesis in spite of C.P. Snow and the two cultures. I think that those of us who are engaged in any form of medicine have to have a very firm scientific base because that is really the thing that distinguishes our capacities now from the capacities of our forebears a hundred years ago. They were just as clever as we are, or as dim if you prefer that, but they did not have the corpus of scientific knowledge on which we can now draw. But equally there is the whole art of ensuring the application of knowledge and to me it is quite unrealistic to try to separate rigidly these two components. They are both valuable capacities of the human mind and I think it is our duty to see that they are applied to the best advantage.

A final word on hazards: do not be so hypnotised by hazards that you stifle innovation and initiative. Thank you, Mr. Chairman.

EPIDEMIOLOGICAL STUDIES IN THE UK
PETROLEUM INDUSTRY

Dr. M. R. Alderson[1] and Dr. L. Rushton[2]

Division of Epidemiology, Institute of Cancer
Research

INTRODUCTION

Comments have been made on the trends in malignant disease in
Western society, as though an explosion in cancer incidence is
occurring (Epstein, 1974). Modern industry has been accused of
generating an increasing risk of malignant disease; however, there
has been wide variation in the estimates (see Bridbord et al, 1978;
Lancet, 1978; Peto, 1980). A number of industries have considered
ways in which the health of their employees could be periodically
reviewed; there has been a general move towards studies of the
patterns of mortality in the various industries. Such studies are
seen as one approach to ascertaining whether there is variation that
requires detailed follow-up. Support for this move has come from
international organizations such as the World Health Organization
(1974) and the EEC (1979). In the UK, the Health and Safety at Work
Act (1974) has explicitly placed the onus on the employer to
identify whether his workers are exposed to risk. It was against
this background that the Institute of Petroleum planned some
epidemiological studies of the industry a number of years ago.*

This article describes some of the findings from 3 inter-related
studies that have been carried out to examine the patterns of
mortality of Oil Refinery and Distribution Centre workers, and
maintenance men in London Transport. (Detailed reports of each of
these 3 studies have been published by the Institute of Petroleum:
Rushton and Alderson, 1980 1982 a; 1982 b.) The principal
conclusion from a case-control study of the incidence of leukaemia
deaths in refinery workers is also discussed. The emphasis in the
present paper is on an overview of the findings from the studies,

1
Chief Medical Statistician, Office of Population
Censuses and Surveys.
2
Now at Thames Polytechnic.

*
Note: A paper by Dr. Alderson - "The Institute's Health Review",
published in the Proceedings of the Institute of Petroleum Annual
Conference, Eastbourne 1976 "Health and Safety in the Oil
Industry" described the inception of these studies.

11

with remarks about the differences in the results from the 3
categories of industry, and some general comments about the conduct
and interpretation of such studies. In addition to the Institute
of Petroleum Reports, contributions are being made of all aspects
of the work to the scientific literature (see Rushton and Alderson,
1980 1981 a; 1981 b).

METHOD

Three historical prospective (ie. retropective) studies have been
carried out by the Institute of Cancer Research with the financial
support of the Institute of Petroleum; they can be considered as
interrelated probes of the health of workers in the petroleum
industry, though not every category of worker was included. The
first study involved eight refineries which had been 'on stream'
at least since the early 1950's (two were associated with
chemical plants, and one produced only bitumen). Entry was restricted
to men who had had a minimum of one years continuous service in the
period 1/1/1950 - 31/12/75. The second study involved all the
bulk distribution centres, numbering some 700 including airport
and blending plants, that belonged to 3 companies in Great Britain;
entry to the study population was restricted as in the refinery
study. The third study involved all maintenance men at London
Transport Executive bus garages and the Chiswick Works, who had had
at least one continuous years service between 1/1/67 - 31/12/75.

For each of these studies details were abstracted from personnel
records at the locations or head offices of the present job of each
worker or the last job for those who had left the industry. For
the refinery study these jobs were coded into one of 13 broad
categories, with a subsidiary code which indicated those on shift
work. For the distribution study, the job titles were coded into
11 broad groups; no information was abstracted on shift work. In the
London Transport study, the jobs were coded into 20 broad groups.
For all men in the refinery study, their present or last known
address was coded; for the distribution and London Transport
studies the address of the location of work was used.

All studies excluded women, as there were only comparatively few
employed in the period under study; men who had spent the greater
proportion of their career abroad - where this could be
identified - were also excluded.

The standard approach was used for identifying the study population,
tracing leavers, calculating person-years at risk, and estimating
expected deaths based on national mortality rates, and the difference
between the Observed (O) and Expected (E) deaths. This was examined
by calculating a significance level under the hypothesis that the
observed deaths are drawn from a Poisson distribution with mean
equal to the externally calculated expected deaths. The comparison
population used was the population of men in England and Wales for
the English and Welsh refineries, and Scotland for the Scottish
refineries. The comparison population for the distribution and
London Transport studies was the population of England and Wales.

For the refinery study, a correction for geographical variation in mortality was made using a simple method based on the Standard Regions in which each of the refineries was located. This was not done for the distribution study, as it would have reduced the numbers for each subset of the analysis too greatly; it was not done for the London Transport study due to lack of appropriate comparison data for Greater London.

The causes of death were grouped into 180 categories, chiefly determined by the availability of data - 30 malignant causes from Case et al (1976) and the A list of 150 causes of death for the non-malignant causes.

The analysis involved two quite appropriate approaches. First, substantial differences between the O and E for the main comparisons (eg. the 8 refineries) were identified 'informally' as those for which the significance level was 0.01. Detailed examination was then carried out (generally) for the results, especially in the refinery study, taking into account the available classifying variables. The significance levels were used as guidelines to increase the magnitude and direction of the variation in mortality and the search for consistency in the findings. Quite different was the second approach, in which the results were used to cross-check the findings from the comparable studies reported in the literature.

In an attempt to investigate the possible effect of benzene exposure in the refinery workers, all deaths with a mention of leukaemia on the death certificate were identified. The potential exposure ofthese subjects was obtained from a 'blind' review of the personnel records of each man. Similar data were obtained for two sets of controls, selected from the entire refinery populations. In one set the three controls were selected by matching on refinery, year of birth, and length of service; the second set were matched only on refinery and year of birth. No information was available on measured benzene exposure for these workers - the available records on job history were used, by a working party of industry staff, to classify each man in the study into potential exposure of one of three arbitrary levels.

RESULTS

There were 34,781 men eligible for the refinery study; 12,525 were still in employment on 31/12/75; 17,078 had left employment but were still alive; 4,406 had died including 3 for whom the cause of death was not available; 679 had emigrated; 73 men could not be traced. In the distribution study, 23,358 men were eligible: 8,470 were still in employment on 31/12/75; 10,904 had left but were alive on 31/12/75; 3,926 had died, including 23 for whom the cause of death was not available; 6 had emigrated; 52 were untraced. In the London Transport study, 8,684 men were eligible 4,671 were still in employment on 31/12/75; 3,102 had left but were still alive on 31/12/75; 705 had died, including 4 for whom the cause of death was not available; 12 had emigrated; 194 were untraced.

None of the men who could not be traced were included in the analyses of any part of the studies. The average length of follow-up was 16.6 years for the refinery study, 17.1 years for the distribution study, but in the London Transport study it was only 5.9 years.

Tables 1 - 3 present the basic comparisons of observed and expected deaths for all causes, all neoplasms, specific neoplasms, and other causes of death for the 3 study populations. This is only a small proportion of the complete analysis of the material, but has been restricted to either those conditions responsible for more than 25 deaths, or where there were at least twice as many observed as expected deaths.

Table 4 contrasts the comparison of O and E for leukaemia from the main refinery study with the results from the specific case-control study of decedents where the certified cause of death included the mention of leukaemia. The results from this latter study are shown for two methods of analysis: i) using the method of Pike and Morrow (1970), and ii) fitting a logistic model. A significant difference in the relative risk of leukaemia in those subjects potentially exposed to benzene is shown when lengths of service is taken into account, either in the matching of cases and controls, or in the analysis.

DISCUSSION

Before attempting to interpret the results of a study such as these, it is important to consider a number of points that bear on the validity of the material: i) Were all the employees initially identified? ii) what are the subtle processes of selection into and out of the industry that might distort the mortality patterns? iii) Did the use of the final job for leavers and retirees prevent the direct examination of jobs that had actually influenced health? (iv) Did the job coding lead to a small group of men with a specific hazard being lost in a broader category? v) Have all the deaths or migrants been correctly identified? (vi) Were there external factors outside the work environment that influenced the health of these workers to a greater extent than the general population (put the other way round, was the use of national mortality rates - adjusted for age, sex, calendar period, and region - a satisfactory approach)? vii) What is the validity of the mortality data and are the certificates for these groups of workers biased? viii) Did the use of national mortality rates provide a suitable estimate of the expected deaths, even when adjusted for region?

These points have been discussed in the full reports to these studies; it is considered that they do not prevent conclusions being cautiously drawn from the study.

The overall level of mortality was reduced in all 3 studies compared with the national standard, giving O/E ratios of 0.84, 0.85 and 0.84. This consistency and reduced ratio is typical of many similar

studies and has been referred to as the 'Healthy Worker Effect'
(Fox and Collier, 1976). It is chiefly a reflection of the higher
levels of mortality that exist in the general population with
inclusion of those who are not fit for work.

As far as the non-malignant causes of death were concerned, most of
those causes for which there were large numbers of deaths had
consistently reduced observed deaths. For example, for Vascular
Diseases of the Central Nervous System (Cerebrovascular disease) the
ratios of O/E were 0.90, 0.87, and 0.95 in the three studies; for
Pneumonia, the ratios were 0.86, 0.72, and 0.87; for Bronchitis,
they were 0.64, 0.67, and 0.77. For Arteriosclerotic and Degenera-
tive (i.e. Ischaemic) Disease of the Heart, the ratios for the
Refinery and London Transport studies were 0.90 and 0.88; however,
for the Distribution workers, a rather different figure of 0.99 was
obtained.

The results for ischaemic heart disease for the distribution
workers seems out of line with the other results for these three
studies and in relation to other studies in the literature. Though
the overall mortality for this cause is not higher than in the
general population, it is not reduced as one would expect from the
other results. This prompted a particular scrutiny of sub-groups
of the distribution centre workforce to check if there was any
evidence of variation in the pattern of mortality. Several excesses
of observed deaths were found in sub-groups of the distribution
population (see Rushton and Alderson, 1982,a). Another point of
interest is the consideration of smoking related diseases (e.g. lung
cancer, ischaemic heart disease, and bronchitis). Lung cancer was
reduced in the refinery and distribution workers, but not in the
London Transport men. Bronchitis was reduced in all 3 studies
(O/E of 0.64, 0.67, and 0.77). It was suggested that the reduced
risk of lung cancer, ischaemic heart disease, and bronchitis in some
of these workers might be due to reduced smoking compared with the
general population. In contrast, for distribution workers,
ischaemic heart disease was not reduced and no obvious explanation
could be provided; it was suggested that further study of this
issue was required.

There was a significant excess of deaths from accidental fire and
explosion for all refinery employees - though based on only 12
deaths, this result was compatible with known past experience in the
locations and provided some support for the validity of the results.

When sub-groups of the various populations were examined, there were
a number of excesses found for various non-malignant causes of death.
In the refinery workers, general manual labourers were found to have
raised risk of Hypertensive Disease, Disease of the Arteries,
Pneumonia, Bronchitis, Peptic Ulcer, and Accidental Poisoning. In
the general manual workers in the distribution study, there were
excesses of Cerebrovascular Disease, Other Diseases of the
Circulatory System, and Asthma; the security men had an excess of
Bronchitis. These differences would have been reduced if rates had
been available for social class adjustment of expected mortality.

There was no evidence in the maintenance men in London Transport of any non-malignant disease being increased as a result of occupation.

The total number of deaths from all neoplasms was significantly reduced in both the refinery and distribution studies, whilst there was a non-significant reduction for London Transport (O/E = 0.95, P = 0.46). Lung cancer showed a highly significant reduction in both the refinery and distribution workers, and this was the reason for the reduced levels of all neoplasms. The lack of difference between the observed and expected deaths from lung cancer in the London Transport study is partly a reflection of the use of national mortality rates; correction for Greater London would have increased the expected figure to give a ratio of O/E = 0.87. There was an excess of Lung Cancer in welders and Cancer of the Bladder in painters - both results compatible with other reported studies.

The refinery workers showed a significant excess overall of cancer of the nasal cavities and sinuses (O = 7, E = 3.1), and of melanoma (O = 14, E = 6.5) (the latter at two refineries in particular). Interpretation of these results is difficult due to the small number of deaths involved. There were no excesses from cancer of the nasal cavities and sinuses or melanoma in the Distribution or London Transport studies. No evidence was found in any of the three studies of raised mortality from epithelioma or scrotal cancer, which have been shown in previous studies to be associated with contact with mineral oil.

Of 5 recent studies on refinery workers, 4 have shown increases of oesophageal cancer, 3 of stomach cancer, and 2 of colo-rectal cancer.* Deaths from cancer of the Oesophagus, Stomach, Intestines, and Rectum were slightly raised overall in the present Refinery study (O = 346, E = 328.6). When sub-groups of the data were examined, increased mortality from cancer of the gastro-intestinal tract was evident in 4 refineries, although different year of entry cohorts and job groups were affected and no location was consistently high for all subsites within the gastro-intestinal tract. Unlike the refinery study, the Distribution and London Transport workers showed no evidence of raised mortality patterns from cancer of the gastro-intestinal tract.

In the Refinery study there was a slight deficit overall from leukaemia (O = 30; E = 31.96). In the Distribution study there were more deaths observed than expected from all Neoplasms of the Lymphatic and Haematopoietic System (O/E = 1.08), but none of the specific forms of neoplasm in this group showed a significant increase overall. When a detailed examination was carried out, raised mortality patterns were found in some groups of the distribution centre populations, although they were based upon small numbers of deaths.

---

*See paper following by Neill. K. Weaver.

The case-control study is difficult to interpret. It must be emphasised that misclassification of the benzene exposure or disease states could produce spuriously high or low estimates of a benzene association. In particular, the lack of documented data on environmental benzene exposure over the years covered by the study meant that knowledge of the refineries and specific plants depended upon the memories of the occupational hygienists in the working party. This necessitated a fairly crude ordinal classification of exposure to benzene, with no attempt to quantify the degree of exposure for different jobs. Taking the above points into consideration, and bearing in mind the apparent absence of an overall increase in deaths from leukaemia, if there was a benzene effect in these men it could only have affected a very small proportion of the men within the total refinery workforce.

Reporting the mortality patterns of workers involved in particular plants is potentially much more disturbing than a study of the general population. Where results show an excess of deaths from any particular cause, this is likely to be interpreted as a causal relationship and highlighted in any further discussions. The individuals at work may consider that the relationship has been shown despite: (a) the general limitations of this type of study, and (b) the specific point that the study was not primarily initiated to test a specific hypothesis. The other side of the coin is that one cannot prove a negative-absence of a significant result may be due to small numbers or limited length of follow-up.

One specific problem is that of assessing the degree of (increased) risk from an individual cause of death, given favourable 'all cause' mortality. The approach selected here has been the conventional one of considering the ratios of O/E above 1.00 to be increased. Where these have been found the overall patterns of mortality have been examined and the inter-relationships of the associations with specific refineries, specific jobs within the refineries or distribution centres, duration of work in the industry, and latent interval probed.

One important point to emphasise about the present report is that it relates to a mixture of workers with varying length of work in the petroleum industry; some will have worked for many years before retiring in the early 1950's, others will have only worked in the post-Second World War period before leaving for other jobs or retiring (if they had been initially recruited into the industry after work in other fields including perhaps military service), while just over one-third of the men were still in employment in the industry. The life-time occupational work history will be very different for these sub-groups of the total study population; even considering only work environment in the refineries this will have been very different for those who worked from the 1930's onwards compared with men joining one of the companies in the 1960's.

A rather different issue is the need to remember that each of the 8 refineries will have had a quite different range of plants and processes in operation; the source of crude oil may vary over time

for one refinery and at one point in time between different
refineries. The separate activities from refinery to refinery, or
from one part of a site to another will mean that the potential
chemical exposure may vary widely for the general category "plant
operatives". Thus differences in reported patterns of mortality of
refinery workers may not be solely due to non-occupational and
chance findings; they may reflect unquantified variation in work
environment. This issue applies to a lesser degree to the men in
the distribution centres or working in different garages for London
Transport. Epidemiological studies in the oil industry are likely
to be beset with problems of mixed exposure during a lifetime of
work, and the small numbers of men who have worked on specific
plants.

It would be most helpful to obtain reactions from the industry to
these studies; do representatives consider that the results have
been of value (have they identified issues requiring further
exploration, have they missed some hazards, have they created or
allayed concern about the health of employees). As an example of
resolving concern, the UK results provide no evidence of a hazard
from brain tumours, such as created interest in the Gulf of Texas
and the US in general (Alderson and Rushton, 1982). As a further
example the excess deaths from melanoma (based on very small numbers)
could not be explained; cross-check of full personnel records at two
locations with 'statistically significant' results showed no common
factor in the men involved. Further follow-up of workers in the
refineries concerned will be complimented by a case-control study of
melanoma in the general population.

Table 1.   OBSERVED AND EXPECTED DEATHS FROM SELECTED
CAUSES OCCURRING IN 34,781 MALE REFINERY
WORKERS IN THE UK IN THE PERIOD 1950-75

CAUSE (IN ICD ORDER)	Observed Deaths	Expected Deaths	O/E	P
All Causes	4406	5259.9	0.84	<.00001
TB Respiratory System	25	63.4	0.39	<.0001
All Neoplasms	1147	1286.4	0.89	.00006
Ca Oesophagus	37	32.4	1.14	.2324
Ca Stomach	167	160.9	1.04	.3143
Ca Intestines	84	78.9	1.07	.2963
Ca Rectum	58	56.4	1.03	.4348
Ca Pancreas	50	51.5	0.97	.4509
Ca Lung and Pleura	416	532.7	0.78	<.00001
Ca Nasal Cavities and Sinus	7	3.1	2.24	.0399
Melanoma	14	6.5	2.16	.0069
Ca Prostate	47	45.9	1.02	.4533
Ca Bladder	34	44.2	0.77	.0669
Ca Brain	36	44.8	0.80	.1054
Leukaemia	30	32.0	0.94	.4089
Vascular Lesions of CNS	408	452.0	0.90	.0192
Chronic Rheumatic Heart Disease	47	75.0	0.63	.0004
Arteriosclerotic and Degenerative (Ischaemic) Heart Disease	1428	1589.7	0.90	.00003
Other Diseases of Heart	85	113.5	0.75	.0038
Hypertensive Disease	76	105.6	0.72	.0020
Disease of Arteries	92	89.3	1.03	.4023
Other Circulatory Disease	39	38.0	1.03	.4557
Pneumonia	157	183.2	0.86	.0264
Bronchitis	253	393.2	0.64	<.0001
Peptic Ulcer	50	54.3	0.92	.3099
Nephritis and Nephrosis	36	42.3	0.85	.1875
Motor Vehicle Accidents	110	101.7	1.08	.2061
Accidental Fire and Explosion	12	6.0	2.01	.0196
Suicide	62	89.1	0.70	.0015

Table 2. OBSERVED AND EXPECTED DEATHS FROM SELECTED
CAUSES OCCURRING IN 23,258 MALE DISTRIBUTION
CENTRE WORKERS IN THE UK IN THE PERIOD 1950-75

CAUSE (IN ICD ORDER)	Observed Deaths	Expected Deaths	O/E	P
All causes	3926	4632.1	0.85	< 0.0001
All Neoplasms	1002	1156.7	0.87	< 0.0001
Ca Stomach	123	144.5	0.85	0.08
Ca Intestines	57	71.7	0.79	0.04
Ca Rectum	57	53.6	1.06	0.34
Ca Pancreas	39	46.9	0.83	0.14
Ca Lung	384	482.8	0.80	< 0.0001
Ca Prostate	53	48.6	1.09	0.28
Ca Urinary Bladder	32	42.8	0.75	0.05
Ca Brain	39	36.5	1.07	0.36
Leukaemia	28	26.8	1.04	0.44
Vascular Lesions of CNS	374	430.7	0.87	0.01
Chronic Rheumatic Heart Disease	31	59.4	0.52	< 0.0001
Arteriosclerotic and Degenerative (Ischaemic) Heart Disease	1377	1384.2	0.99	0.84
Other Diseases of Heart	71	110.7	0.64	< 0.001
Hypertensive Disease	65	91.9	0.71	< 0.01
Disease of Arteries	81	88.9	0.91	0.22
Other Diseases of Circulatory System	36	35.9	1.0	0.52
Pneumonia	133	185.3	0.72	< 0.0001
Bronchitis	240	358.6	0.67	< 0.0001
Peptic Ulcer	40	50.7	0.79	0.07
Motor Vehicle Accidents	75	69.9	1.07	0.29
Suicide	38	67.1	0.57	< 0.0001

Table 3. OBSERVED AND EXPECTED DEATHS FROM SELECTED
CAUSES OCCURRING IN 8,684 MALE MAINTENANCE
WORKERS EMPLOYED BY LONDON TRANSPORT
EXECUTIVE IN THE PERIOD 1967-75

CAUSE (IN ICD ORDER)	Observed Deaths	Expected Deaths	O/E	P
All Causes	705	836.2	0.84	<0.0001
All Neoplasms	216	226.8	0.95	0.46
Ca Stomach	26	25.5	1.02	0.49
Ca Lung	102	101.3	1.01	0.94
Vascular Lesions of CNS	58	74.4	0.78	0.03
Arteriosclerotic and Degenerative (Ischaemic) Heart Disease	244	276.6	0.88	0.05
Pneumonia	28	32.0	0.87	0.27
Bronchitis	50	64.7	0.77	0.04

Table 4。 LEUKAEMIA I N 35,000 OIL REFINERY WORKERS IN THE
UK, 1950-75

(a) Leukaemia as the underlying cause of death in the total study
population:

O = 30, E = 31.96, P = 0.409

(b) Case-control study of benzene exposure for 36 deaths
mentioning leukaemia (i.e. 30 where leukaemia was the
underlying cause plus 6 where leukaemia was also mentioned):-

(i) Pike and Morrow method:

Controls matched on age χ^2, = 2.3; P = 0.13

Controls matched on age and length of service χ^2, = 3.0;
P = 0.08

(ii) 'best fit' logistic model:

Controls matched on age   RR = 2.99 (1.24 - 7.20)

Controls matched on age and length of service RR = 2.33
(0.98 - 5.56)

REFERENCES

M.R. Alderson and L.R. Rushton, Anns New York Acad. Sc., 381, 139-146 (1982).

K. Bridbord, R. Decoufle, J.F. Fraumani et al, Estimates of the fraction of cancer in the United States related to occupational factors, National Institute of Health (1978).

R.A.M. Case, C. Coghill, J.M. Davies et al, Serial Mortality Tables, Neoplastic Disease Volume 1 - England and Wales, 1911-70, Institute of Cancer Research, (1976).

S.S. Epstein, Cancer Res., 34, 2425-2435 (1974).

European Communities, Official Journal of the European Communities, 22, 6-9 (1979).

A.J. Fox and P.F. Collier, Brt. J. of Industrial Medicine, 33, 249-264 (1976).

Health and Safety at Work Act, Chapter 37, HMSO, London (1974).

Lancet, 2, 1238-1240 (1978)

R. Peto, Nature, 284, 297-300 (1980).

M.C. Pike and R.H. Morrow, Brit. J. Prev. Soc. Med., 24, 42-44 (1970).

L.R. Rushton and M.R. Alderson, An epidemiological survey of 8 oil refineries in the UK - final report, Institute of Petroleum, London, 1980a).

L.R. Rushton and M.R. Alderson, Carcinogenesis, 1, 739-743 (1980b)

L.R. Rushton and M.R. Alderson, Brit. J. Industr. Med., 38, 225-234, (1981a)

L.R. Rushton and M.R. Alderson, Brit. J. Cancer, 43, 77-84 (1981b).

L.R. Rushton and M.R. Alderson, An epidemiological survey of oil distribution centres in Great Britain - final report, Institute of Petroleum, London (1982a).

L.R. Rushton and M.R. Alderson, An epidemiological survey of maintenance workers in London Transport Executive bus garages and Chiswich works - final report, Institute of Petroleum, London (1982b).

World Health Organisation, Report EHE/75.1, WHO Geneva (1974).

DISCUSSION

Peter Jones (Institute of Petroleum)

To answer Dr Alderson's question, the Institute certainly feels that these three studies and their results have been most worthwhile. There is nothing so comprehensive on this side of the Atlantic, and it has been a privilege to participate in them and the conveyance of their results to our colleagues in Europe and elsewhere who look to the Institute of Petroleum as a source of information.

As to the value of the results, Dr Alderson referred to "the negative" - but I would comment as to the point of view of an industry seeking the information that epidemiology can give, that the negative is as important to know as the positive, and tells as much. Take for instance the fact that in these three studies there is the negative that there is not one case of scrotal cancer mortality and a deficit in observed deaths from epithelioma (cancer of skin), both of which have been of historic interest. Dr Leese will in his paper report the study which the I.P. has had carried out on animal testing on the skin of mice, and while these tests have to be done there is then the uncertainty of what they mean in terms of humans under very different conditions. The methodical review of 80,000 persons by Drs Alderson and Rushton and their reporting is of particular value in guiding us that our health and hygiene precautions in our codes of practice are correct and are getting through to the people concerned - and that the observations in the animal studies are not necessarily replicated in man.

In a sense too the three studies provide a follow up, with their individual findings enhancing that of the others, for many of the exposures of the three study populations have much that is common. The reporting of subgroups in which deficits are to be found is also of particular significance as where comparable subgroups show some excess. We think the studies of immense importance to the industry and that they will be an example and an encouragement to others.

A REVIEW OF DGMK's CURRENT HEALTH RESEARCH ACTIVITIES

A. Kluge

Department Head, Occupational Medicine, Toxicology,
Industrial Hygiene. Deutsche Gesellschaft für
Mineralölwissenschaft und Kohlechemie E.V. (D.G.M.K.)

I should like to thank the President of the IP very much for his
invitation to present a review of the health research activities of
DGMK (German Society for Petroleum Sciences and Coal Chemistry) on
the occasion of the 1982 Annual Conference.

The health research activities conducted by DGMK since 1978 and our
current and future activities are subject to the so-called
preventive principle of the Federal Government of the Federal
Republic of Germany:

"The underlying principle is not to repair ecological damage but
rather to prevent such damage 'a priori' by means of foresighted
measures. The realization of this principle also requires the
registration and evaluation of such substances that are not in the
centre of topical interest for early identification of potential
hazards."

Amongst others, the following consequences result from this principle
for the German petroleum industry:

1. The preventive principle will increasingly influence existing and
    new technologies for the manufacture and application of
    petroleum and petrochemical products as new toxicological and
    industrial hygiene findings become known, entailing specific
    questions regarding the effects of substances on those employed
    in the industry, on consumers and on the environment.

2. New findings may moreover lead to entirely different toxicological
    evaluations and also for those substances already well known and
    analyzed.

3. Hence the petroleum industry needs to monitor the development of
    the state of knowledge of the effect of substances in order to
    be in a position to provide replies to any specific questions.

4. Scientifically founded arguments are required absolutely for
    this purpose. This means basically that for the manufacturing,
    processing, handling and application sectors of petroleum and
    petroleum products there is the need for determination and
    evaluation of

    - the harmful effect of given substances and products on man,
      animal and plant life

and

- the occurrence of these substances and products on the job and
  in the environment.

The management of DGMK has decided that the problems involved, as far
as they are of importance for the petroleum industry, shall be
handled by DGMK. In its capacity as a neutral and scientifically
recognized society, DGMK has access to the necessary knowledge of
specific substances and industries.

It is readily obvious that with such a difficult subject decision
processes are lengthy. According to the current status of discussions,
the activities of DGMK are concentrating on 5 aspects compiled in
table 1.

Aspect 1 "Pure Substances" comprises the following reports:

1. Research report 174-6:

"Evaluation of Benzene Toxicity in Man and Animals", published in 1980
meeting with great interest at home and abroad.

This report attempted to achieve:

- clarification of diverging ideas on the matter of threshold limit
  values

- covering of the subject, discussed again and again "benzene in
  motor gasoline" on the basis of facts taken from practical
  experience

- inclusion of the DGMK report in a planned publication of the
  German Federal Environment Agency entitled "Air Quality Criteria
  for Benzene".

After protracted discussions of the DGMK findings the Federal
Environmental Agency agreed to include the DGMK benzene report in a
slightly abridged version in its publication.

The essential statements of this report were:

"Despite the fact that a dose-effect relationship has not been
defined, the careful review of the literature has shown that
fluctuations in the blood picture over and above the normal, as
detected by established diagnostic methods, first occur clearly
at benzene concentrations in excess of 20 ppm, in animal experiments,
however, above 31 ppm.

Also for benzene-induced leukaemia it has not been possible to
define a clear dose-effect relationship. With certain reservations
it can be taken from the literature that no benzene-induced
leukaemias have been observed below 50 ppm. Furthermore, no clear
references to benzene-induced leukaemias are available for the
concentration range 60-150 ppm, and from this we may conclude that at
benzene concentrations below 100 ppm no leukaemias occur. Experience
and observations to date have shown that for man benzene has only a

relatively weak leukaemogenic effect. Such a fact could either not be demonstrated at all in animal experiments or only under exceptional test conditions. In addition, the negative results in the Ames test support the conclusion that the mutagenic potential of benzene is extremely low.

To summarize, although there is no guaranteed dose-effect relationship for benzene in man, from the many epidemiological investigations, case studies, animal tests, in vitro tests and chromosome analyses, it is evident that this widely distributed substance exerts only a weak carcinogenic effect above 100 ppm. Furthermore, it is apparent that personnel who work with benzene are obviously adequately protected by the 'Technical Guiding Concentration' (TRK) of 8 ppm."

(Technical guiding concentration values are assigned only to carcinogenic working materials for which MAK-values (maximum concentration values in the workplace) confirmed by toxicological or industrial medicine experience cannot be established at present time. Adherence to technical guiding concentration values is meant to reduce the risk of health hazard but cannot completely eliminate it.)

2. In May of this year we published research report 174-2 on:

"Evaluation of n-Hexane Toxicity in Man and Animals".

Numerous publications exist on the effect of n-hexane, or substances containing n-hexane, on man and animal after, for the first time in Japan in 1964, cases of polyneuropathy were observed among workers who had been exposed to n-hexane over a long period.

As n-hexane is a component of many petroleum products, eg. the special boiling point solvents widely applied in the industry and in motor gasolines, there have been worldwide discussion for some time on revising the currently valid data on maximum n-hexane concentration rates allowed in the workplace.

We therefore considered it to be necessary in the case of this substance also to assess the current status of knowledge of n-hexane and subject it to a critical evaluation. Allow me to report in further detail the outcome of this study:

This study provides detailed information on:

- the industrial application of hydrocarbons containing n-hexane

- the occurence on n-hexane in the environment

- the uptake and biotransformation of n-hexane

- the distribution, elimination and determination of n-hexane and its metabolites and in particular:

- the present scientific and practical knowlege relating to the effect of n-hexane on man and animals.

Main results of this study are:

The occurrence of n-hexane in the environment is due mainly to emissions from the operation of motor cars with internal combustion engines and emissions of industrial solvents. Annual n-hexane emissions in the Federal Republic total roughly 15,000 t (see table 2).

The principal mode of uptake for n-hexane in man is resorption via lungs upon inhalation with respiratory air. Only about 7% of the hexane contained in the respiratory air is retained in the body and metabolized. The metabolites are discharged in the urine.

Biotransformation of n-hexane is effected essentially in the liver. The final product of the metabolic transformation is the toxic metabolite 2,5-hexanedione causing the neurotoxic effect of n-hexane (see table 3).

According to our current knowledge, the identical mechanism must be assumed to exist for biotransformation of n-hexane in man and animal. There is, however, a difference in quantity: While, in the urine of man 2,5-hexanedione was found at the rate of about 20 to 30 times the volume of 2-hexanol, urine analyses in animal tests showed greater quantities of 2-hexanol, the neurotoxic potential of which is essentially lower.

Acute poisoning through n-hexane inhalation leads to a depression of the CNS resulting in a state similar to anaesthesis. A dose-effect relationship cannot clearly be derived from the literature. However, it may be assumed that the concentration of roughly 1,500 to 2,500 ppm of hexane leads to pre-anaesthesia symptoms.

Chronic inhalation of n-hexane may lead to a symmetrical polyneuropathy, mainly affecting the lower parts of the limbs with loss of sensation and, in advanced cases, some muscle weakness. In parallel with these effects electrophysiological disturbances and histological changes in the peripheral nerves can be identified.

Rehabilitation, depending on the degree of harmful effect, takes generally 6-36 months after termination of the exposure.

Poisoning effects after acute and chronic effects are essentially identical in man and animal.

According to experience gained so far, n-hexane has neither a teratogenic or carcinogenic effect on animals, nor there have been any publications of tests regarding mutagenic, carcinogenic or teratogenic effects on man.

When summarizing the results of tests in man taken from the literature one may say in a conservative assessment that polyneuritic changes in the case of chronic effect of n-hexane occur with roughly 100 ppm. However, this dose-effect relationship is not certain as the data on the composition of the working substances, the concentration rates in the workplace and the measurement methods are mostly inaccurate or incomplete. Most studies quote very wide concentration ranges without readily identifying the concentration rates to which the persons examined were actually mainly exposed. Concentration rates are between 10 and 3,000 ppm.

According to current knowledge we may procede on the assumption that the different biotransformation rates of n-hexane in man and animals is the reason why the neurotoxic effect on animal occurs only above 200 ppm while the human organism may react with polyneuropathic symptoms even at 100 ppm. From this may be deducted that, on the one hand, translation of the results with animal cannot readily be effected to man in the case of n-hexane but, on the other hand, that the current fixed time-weighted average concentration of 100 ppm probably cannot protect personnel entirely from possible health hazards.

The result of this study has made obvious that for the establishment of a generally accepted dose-effect relationship further studies, including matched control groups are necessary and precise analyses of the working substances containing n-hexane employed and careful, standardized occupational exposure readings are required.

We feel that this research report on the question of the effect of n-hexane on man and animal provided an essential contribution and represented the necessary factual basis for future decisions. The report was therefore submitted to the MAK-Commission (Commission for Investigation of Health Hazards of Chemical Compounds in the Air; chairman: Prof. Dr. D. Henschler) as this Commission is discussing the current valid maximum concentration of n-hexane within a working area.

3. Another item in Aspect 1 is the following:

For the testing of motor fuels containing methanol the Ministry of Research and Technology has been conducting a large-scale project with roughly 1,000 vehicles since 1980 in which DGMK also plays a major role, amongst other things with the sub-project "Safety and Environmental Aspects in the Handling of Motor Fuels Containing Methanol". This sub-project includes the study "Evaluation of Methanol Toxicity on Man and Animals" which is intended to provide all information on the present status of knowledge of methanol. This study is currently being prepared by us and will be available by autumn.

With project 250 "Measurement of Benzene Exposure in the Handling and Manufacturing of Motor Gasoline" we are currently covering a subject of <u>Aspect 11</u> "Exposure and Analysis".

This project will make available statistically validated data on benzene exposure of man at retail outlets, loading facilities and manufacturing plants. Moreover, it is planned to investigate whether benzene exposure is related to the benzene concentration in motor gasolines. The results should lead to exact information on the actual benzene exposure on the job. Any demands envisaged from legislation on procedures at workplaces and investments possibly resulting from them may be evaluated after our results are available.

This project involved 9 companies who, with their own staff, conducted the exposure measurements after the sampling and analysis methods had been fixed by numerous collaborative tests, which proved that two methods are appropriate for the determination of benzene in air, namely the TENAX- resp. PORAPAK-method and the charcoal-method. Exposure readings were made in the spring, summer, autumn and winter of 1981. Table 4 shows the type and number of exposure measurements.

Measurements on service station attendants and autobahn service stations have already been statistically analyzed, both separately for each parameter influencing the individual and also for combinations of two or three parameters. Table 5 lists all parameters taken into account.

The benzene concentration rates as measured were

- on service station attendants:

  0,003 - 1,1 ppm;   mean value = 0,16 ppm

- on attendants on autobahn service stations:

  0,006 - 1,23 ppm;   mean value = 0,15 ppm

I will mention only the statistical analysis of the readings taken at autobahn service station attendants which gave the following picture:

The parameters

- test week and, by implication, ambient temperature

- weather

- wind

- traffic density

- number of deliveries

- fuel volume delivered

have significant influence on the benzene values.

The final report of this project will be published by autumn.

Another item of Aspect 11 is the project of the Ministry of Research and Technology, which I mentioned before, as it includes in addition

- the development of a suitable method for sampling and analysis of methanol in vapours also containing hydrocarbons, and

- the measurement of methanol exposure in the filling of motor cars with motor fuels containing methanol.

A sampling and analysis method has already been developed, and the exposure readings are planned to be effected by 3 companies this month. The final report of this project will be available also by autumn.

Our tests seeking an exact evaluation of the cadmium emissions in the Federal Republic from the burning of fuel oil and heating oil may also be allocated to Aspect 11.

First of all a suitable analysis method had to be developed for the identification of cadmium in heating oils, as the cadmium content is only roughly 10 ppm.

At present collaborative tests are being conducted for the determination of the cadmium content in heating oils of different origin and in heating oil residues for subsequent preparation of an accurate cadmium emission balance sheet. Such balance sheets are of importance for the continuing discussion in the Federal Republic of Germany on the effects of cadmium.

Aspect 111 "Prospective Epidemiological Studies" is most difficult and time-consuming. It is currently under discussion in the DGMK member companies. These discussions are based on the following considerations:

So far, medical research of effects has been concerned with studies of harmful effects that have already occurred with substances. In future, however, the preventive evaluation of potential risks will increase in importance, as the industry must bear in mind the preventive principle already mentioned. This means that both well known and entirely new techniques must be evaluated against the occurrence of health hazards. This requires epidemiological tests.

For such tests the individual works medical officer in the majority of cases does not have an adequate number of personnel at his disposal and also, by implication, not the data required for a statistically validative evaluation of the test results. It is only by the registration and analysis of an adequate number of epidemiological test results that he will be in a position to investigate the reasons of any health hazards of occupational origin.

The members of the DGMK committee "Occupational Medicine, Toxicology, Industrial Hygiene" held the following view:

"There is an undisputed need for development and documentation of epidemiological data in the research sector of DGMK. These are essential for the toxicological assessment of chemical effects on man and the environment.

Epidemiological tests should therefore be conducted jointly by the German petroleum industry. The task of DGMK would, amongst other things, be to coordinate the epidemiological tests at the DGMK member companies in order to be able to draw on an adequate number of personnel groups."

Therefore, the DGMK committee has suggested to start first of all with a minimum program involving only one working substance or refinery stream, having due consideration to the studies conducted so far by IP and API.

One working group have already been set up which is currently preparing a study for the planning, conducting and evaluating of epidemiological tests of personnel in the petroleum industry.

This study will be available by September and then be submitted to the managements of the DGMK member companies for review and decision.

The DGMK project 160 entitled "Medical Evaluation of the Components of Cutting Fluids" belongs to Aspect 1V "Industrial Hygiene Related to Commercial Products".

Water-miscible and non-water-miscible mineral oils based cutting fluids are used in great quantities for metal working. As it is known from experience that health hazards have been frequently observed in their application the DGMK committee has suggested a register of all cutting fluid components used in practice for subsequent toxicological evaluation. This evaluation would then show which components offer health hazards, if any, and which must therefore no longer be used.

In cooperation with cutting fluid manufacturers great efforts have been made to compile a comprehensive list of substances which contains virtually all components currently used for the manufacturing of cutting fluids. It is a very long list with 143 substances divided into 34 classes, such as polyglycol ethers, aliphatic amines, natural fatty acids, phenols, phosphoric acid esters, chlorinated organic compounds, etc.

This list was made available last month to the MAK-Commission with the approval of the cutting fluid manufacturer, as this commission had informed DGMK recently that they had been asked by several organisations to prepare suitable threshold limit values for cutting fluids. This, however, requires that the components used for the manufacturing of cutting fluids are officially disclosed, as a limitation to general comments would be counterproductive for the purpose of work protection.

The MAK-Commission has appreciated the receipt of our list as it represents a factual basis for its future work. The toxicological evaluation of these 143 substances is planned in cooperation between DGMK and the MAK-Commission. This will no doubt take a long time due to the high number of substances.

In this context it has to be mentioned that German authorities have pointed out health hazards in the handling of cutting fluids and are already conducting tests of their own.

Further Aspect lV activities are the following:

In the years 1977 to 1982 DGMK has been involved in research projects of the Federal Environmental Agency on the effect and emission of polycyclic aromatic hydrocarbons (PAH).

The aim of the DGMK projects was to develop suitable collection and analytical methods for the determination of PAH in automotive exhaust gas and in the flue gases of commercial oil-fired central heating units with high-pressure jet burners and vaporizing pot burners. Exact and reproducible data for the various PAH mass emissions were obtained.

Considering Aspect V I would like to mention the DGMK projects

- Measurement and determination of hydrocarbon emissions during storage and transportation of gasoline
- Hydrocarbon emissions during filling of tank trucks with gasoline
- Test unit to reduce hydrocarbon emissions during filling of tank trucks with gasoline

which were published during 1977-1980.

In another project we investigated the effects of hydrocarbons on plants. The report "Effects of Hydrocarbons on Higher Terrestrial Plants" which was published end of 1978 gave a comprehensive review of the effects of a large variety of hydrocarbons which are listed in table 6.

The abstract of this study stated:

"Plants are injured by atmospheric hydrocarbons only if the hydrocarbon concentration considerably surpasses the actual concentrations measured in the atmosphere.

According to published research work ethylene should be considered as a special case as it is produced by plants themselves. At low concentrations ethylene may develop multiple effects in plants, eg. effects on plant growth which are generally limited to so-called invisible damage, such as growth retardations."

So much for my review. Unfortunately, I was in a position to present our activities in a summary only. Nevertheless, I very much hope some of the items I have mentioned may have been of interest to you.

Thank you for your attention.

Note on nomenclature: The terms "PAH – polycyclic aromatic hydrocarbons" and "PCA – polycyclic aromatics" used later in these papers are synonymous. In an earlier period they have been referred to as "polynuclear aromatics (PNA's)".

BASIC ASPECTS OF DGMK'S HEALTH RESEARCH ACTIVITIES

1. Pure Substances

    Critical assessment and presentation of the status of knowledge on toxicological and occupational
    medicine experience in the handling of pure substances

11. Exposure and Analysis

    – Development of suitable standardized analysis techniques for accurate identification of the
      substance in question itself and in mixture of substances
    – Conducting of exact exposure measurements on the job
    – Development of suitable standardized methods for identification of the metabolites of the
      substance in question in body fluids (Biological Monitoring)

111. Prospective Epidemiological Studies

    Application of the results from 1 and 11 in the conducting of epidemiological tests

IV. Industrial Hygiene related to Commercial Products

    – Toxicological properties of commercial products
    – Analysis of the substances that can be formed in the application of commercial products

V. Environmental Problems

    – Emission of hydrocarbons
    – Effect of hydrocarbons on plants

TABLE 1

ESTIMATED n-HEXANE EMISSIONS

Storage, Transport and Distribution of Motor Gasoline	ABT.	948	TO/YEAR
Operation of Gasoline Passenger Cars			
– Displacement and Breathing Losses	"	887	
– Exhaust	"	7,730	
Industrial Solvents	"	6,000	
	ABT.	15,565	TO/YEAR

TABLE 2

39

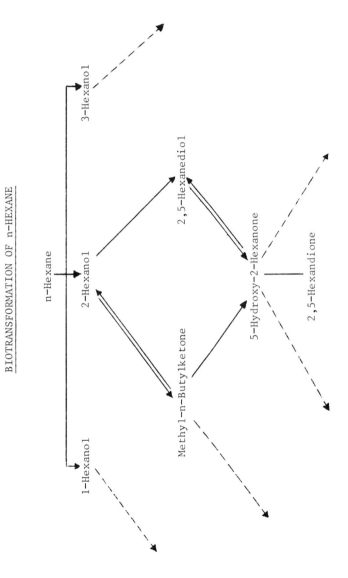

BIOTRANSFORMATION OF n-HEXANE

n-Hexane

3-Hexanol

2-Hexanol

1-Hexanol

2,5-Hexanediol

5-Hydroxy-2-Hexanone

Methyl-n-Butylketone

2,5-Hexandione

TABLE 3

PERSONAL EXPOSURE TO ATMOSPHERIC BENZENE ASSOCIATED WITH DISTRIBUTION AND MANUFACTURING
OF MOTOR GASOLINE

Locations	Personal	Number of Measurements
SERVICE STATIONS		
- in towns	Attendants	256
- at motorways	"	97
BULK LOADING FACILITIES		
- truck loading	Operator	37
- rail car loading	"	80
- marine loading	Deck Crew and Pier Operatives	65
TRUCKS	Driver	57
REFINERIES		
- production of motor gasoline	Operator	122
- production of motor gasoline and streams with < 10% benzene	"	165
- production of benzene and streams with > 10% benzene	"	130
		1,009

TABLE 4

BENZENE EXPOSURE MEASUREMENTS

– PARAMETERS CONSIDERED –

Test Week   Number of Refuelings

Sampling Time   Refueled Gasoline (in litres)

Weather   Benzene Content of Motor Gasoline

Wind   Benzene Quantity in Refueled
                                   Gasoline (in kg)

Density of Traffic
in front of the Service   Analysis Method
Station

TABLE 5

EFFECTS OF HYDROCARBONS ON PLANTS

Referred published studies include the following hydrocarbons:

Paraffines
_____

Methane
Ethane
Propane
n-Butane
i-Butane
n-Pentane
i-Pentane
n-Hexane

Cycloparaffines
_____

Cyclopentane
Cyclohexane

Monoolefines
_____

Ethylene
Propylene
1-Butene
2-Butene
1-Hexene

Diolefines
_____

Allene (Propadiene)
1,3-Butadiene
Isoprene

Cyclomonoolefines
_____

Cyclopentene
Cyclohexene

Acetylenes
_____

Acetylene
Methylacethylene (Propin)
1-Butine

Aromatic Hydrocarbons
_____

Benzene
Toluene
Xylenes
1,2,3 -Trimethylbenzene
1,2,4 -Trimethylbenzene
1,3,5 -Trimethylbenzene
1,2,4,5-Tetramethylbenzene

TABLE 6

U.S. OIL INDUSTRY EPIDEMIOLOGY

Dr. N.K. Weaver

Medicine and Biological Science Director,
American Petroleum Institute, Washington, DC.

ABSTRACT

Refineries are suitable sites for epidemiology studies designed to
assess possible petroleum-related health disorders since the workers
form a relatively stable population with potential exposure to a
range of crude oils, intermediates and products. The American
Petroleum Institute (API) first undertook such a study during the
period 1950-1956, focusing on cancer experience in the industry
with particular emphasis on tumor occurrence in refinery workers.
This was followed by a retrospective study of mortality in
refinery workers (1962-1976), and a prospective surveillance program
covering mortality, morbidity and tumor incidence in refinery,
petrochemical and laboratory employees (1977-1981). The findings of
these API epidemiology studies are presented. Perspective is added
through comparative analysis with published reports of other
investigations which were designed to generate hypotheses, to
provide surveillance or to test hypotheses concerning mortality
(especially cancer-related mortality) in refinery workers.

The status of new API epidemiology programs dealing with data bases
and methodologies for evaluation of possible reproductive effects
and the "healthy worker effect" is also presented.

INTRODUCTION

Mr. Chairman, Ladies and Gentlemen: It is indeed, both a
privilege and a pleasure to participate in this 1982 conference of
the Institute of Petroleum. In the presence of such distinguished
company, I am mindful of the honor inherent in an opportunity to
address leaders in the petroleum industry in the United Kingdom and
your guests from various parts of the world. On this prestigious
occasion my comments will draw heavily upon the activities and
experiences of medical colleagues from member companies of the
American Petroleum Institute.

I also thank our hosts for the informal meetings and get-to-gethers
arranged for further discussion of topics of occupational and
environmental interest. From my American Petroleum Institute (API)
staff position, I have long enjoyed a most cordial and constructive
working relationship with Mr. Peter Jones and, more recently,

Mr. Stan Parkinson of the Institute of Petroleum staff. It is nice
to supplement our usual flow of correspondence, and occasional phone
calls with a direct, face-to-face exchange! And let me assure you
that we in the Medicine and Biological Science Department of API are
always pleased and flattered to welcome IP representatives to our
meetings. Dr. B.W. Duck, Chairman of the IP Advisory Committee on
Health, Dr. W.L.B. Leese, immediate past Chairman of ACH, Dr. Ian
Holle and Dr. A.H. Pickering (to name a few) have participated in
certain of our recent meetings in a liaison capacity between our
respective organizations - a role they have filled ably and well.

It is noteworthy that the Institute of Petroleum has designated
Heath and Hazards in a Changing Oil Scene as the theme of this
conference. The selection of this topic points up its importance
in the eyes of petroleum company managers and lends encouragement
to us who labor in the occupational health field. The international
scope of this opening session demonstrates that health and safety
interests and concerns transcend national, regional and
continental boundaries. They attract a great deal of attention on
the part of labor, management, government, environmental groups and
news media, especially in the U.S. (For those of us who have
experienced the traditional practice of occupational medicine in a
quiet, conservative setting, this transformation into the limelight
has been rather disconcerting.)

In presenting selected aspects of our epidemiology program, I shall
endeavor to convey some of the rationale surrounding these
investigations. I hope that such insight may prove worthwhile in
providing meaning and perspective to the research findings. First,
I shall describe three relatively mature API-sponsored epidemiologic
investigations. Next, I shall discuss what is currently a very
prominent epidemiologic issue - the incidence of cancer as a
cause of mortality in refinery workers. Then, two new API
epidemiologic studies will be briefly described. And finally, with
the aforementioned examples as a basis, I shall address in a rather
philosophical vein the state-of-the-art of epidemiology as it exists
in the U.S. industrial setting today.

THREE API EPIDEMIOLOGICAL STUDIES

The three studies selected for presentation in this discussion are:

1. MC-1 An Epidemiological Study of Cancer among Employees in the
    American Petroleum Industry.(1)

2. OH-1 A Mortality Study of Petroleum Refinery Workers.
    I.R. Tabershaw, M.D., Tabershaw/Cooper Assoc., 1975; S.D. Kaplan,
    M.D., SRI International, 1982. (2)(3)

3. OH-11 A Prospective Mortality, Morbidity and Tumor Registry
    Study in Petroleum Workers.
    D. Schottenfeld, M.D., Memorial Sloan-Kettering Cancer Center,
    1981. (4)(5)

For ease of identification, they will be referred to as the
"Kettering","Tabershaw" and "MSKCC" studies.

THE "KETTERING" STUDY

The first study is predominantly of historical interest. An API Medical Advisory Committee (the predecessor of the present Medicine and Biological Science Department) was formally established in 1945, and planning for a study of the occurrence of cancer among employees in the American petroleum industry must have started soon thereafter, since the collection of data was taking place in 1949. Data collection continued during the period 1950-1956, with subsequent analyses, and, finally, issuance of a report in 1958. We take some pride in anticipating events in the overall research planning for the Department, and this study was well ahead of the times! There were certainly no publicly expressed concerns - no headlines! - and no regulatory initiatives dealing with occupational cancer in the 1940's and 1950's. What motivated the initiation of such a study in 1949? The decision was a sound and responsible action on the part of industry medical scientists and managers who recognized a carcinogenic potential in petroleum. Interestingly, development of this research involved quite extensive interaction between U.S. and British Industry medical scientists and investigators.

The induction of skin tumors by repeated application of crude petroleum to rodents was first demonstrated by Leitsch in England in 1922 (6), a finding which was subsequently corroborated by additional studies in the U.K. (7)(8), the USSR (9) and the U.S. (10)(11). As you are well aware, certain polycyclic aromatic hydrocarbons found in the higher boiling fractions were identified as carcinogens in animal experiments. The recognition of this potential carcinogenic activity led the Medical Advisory Committee of the API to recommend a comprehensive study of the problem. The Kettering Laboratory of the University of Cincinnati was authorized to undertake investigations which included chemical analyses and biological tests of various fractions beginning in 1948. They started the epidemiological study the following year.

On this side of the Atlantic, the British National Research Council established a "Special Committee on the Carcinogenic Action of Mineral Oils" in 1948 which also undertook chemical and biological tests of crude petroleum samples and extracts (12). While the British did not include epidemiological or clinical aspects, the U.S. and British analytical and biotest programs were coordinated. The U.S. focused on fractions derived from catalytic cracking, and the British studies on those from thermal cracking. Dr. Phair, the principal investigator for the API cancer incidence study, and others from the Kettering Laboratory visited U.K. researchers on a number of occasions, and the teams maintained contact during the course of their respective investigations.

In reviewing the project file, it is evident that Dr. Phair experienced the same concerns and frustrations about completeness and accuracy of reporting that beset epidemiologists today. Seventeen companies distributed across the U.S. participated. While tumor collection process was industry-wide in scope, the majority of reports emanated from refineries, apparently for logistical reasons. In all, 2,108 tumors were reported, which Dr. Phair classified by histological type and site of origin. A serious drawback was lack of adequate control information, due to the fact that the study was well in advance of the general population studies. The U.S. Public Health Service had published cancer data for 10 major cities, which were used as the "expected" occurrence for comparison. After completion of analyses, Dr. Phair concluded that "no significant differences were found (between the petroleum workers and the "expected" occurrence for comparison)..." He emphasized, however, that due to the limited and very small sample of cases, "this finding must be interpreted with unusual caution".

THE "TABERSHAW" STUDY

The "Tabershaw" study is a retrospective, or historical prospective, investigation of mortality in petroleum refinery workers. Since API's "Tabershaw" and "MSKCC" studies, and indeed the IP study conducted by Professor Alderson (13)(14), focus on refinery populations, the rationale for selecting this component of petroleum operations warrants comment.

For a number of reasons, refinery workers are particularly important as subjects for study by epidemiologists concerned with assessing the potential effects of the petroleum industry on human health. Employees engaged in petroleum refining are concentrated in better defined populations than are workers in the production, transportation and marketing segments of the industry. Refineries provide continuity in employment; hence, the worker populations are relatively stable. Since refining operations involve the processing of crude oil to a full line of products, the refinery worker is potentially exposed to the complete spectrum of petroleum hydrocarbons and intermediates; and to the attendant emissions into the atmosphere, effluents into aqueous systems and solids for waste disposal. While circumstances may vary, refinery employees are apt to be exposed to these materials at an earlier date, at higher concentrations and for longer periods of time than populations in other parts of the industry, amongst its customers, in communities adjacent to refineries, or in the general public. Furthermore, for an industry quite lacking in obvious and severe health hazards, workmen in many refineries are provided rather intensive and thorough programs of health surveillance.

The "Tabershaw" study was conducted in 17 refineries selected to be representative of the U.S. industry as a whole with respect to size, geographical distribution, crude and products, processing methods and other factors (2). The study group consisted of 20,131 male hourly employees, not clerical, who had worked in the refinery for at least one year between 1962 and 1971. There were 1,194 deaths, the follow-up being more than 99% complete. Subsequently, the study was extended an additional 5 years (through 1976), at which time the number of deaths had increased to 2,237, with follow-up 97% complete.

The mortality experience of refinery workers throughout this study has been quite favorable. At completion of the second phase in 1976 the total number of deaths was 76% of what would have been expected, a statistically significant reduction. (3) Specific cause of death for which there were significantly fewer deaths than expected included: all infective and parasitic diseases (including tuberculosis); malignant neoplasms as a group, and specifically, cancer of the respiratory system, lung, larynx and bladder; diseases of the circulatory system overall, and arterosclerotic (ischaemic) heart disease in particular; all respiratory diseases, including pneumonia and emphyzema; and diseases of the digestive system in general.

The only statistically significant excess mortality occurred for cancer of "other lymphatic tissue", based on 24 deaths 13 of which were due to multiple myeloma.

The following statement by Dr. William Gaffey, the biostatistician who analyzed the first phase of this study, appropriately reflects his view of the overall favorable mortality experience: "If I were a young person seeking employment in a situation with a good chance to achieve longevity, I could rationally choose to work in a refinery".

Originally intended as a pilot or a feasibility study, the power of the analyses exceeded expectations with respect to the significance of findings, and it was decided to extend the study by adding 5-year follow-up intervals rather than to expand the size of the study population. In this context, the results cited may be viewed as interim rather than final. Consequently, the study has never been formally published, although more than 1,000 copies of the report have been made available to interested parties in the scientific community upon request. We are planning a new 5-year follow-up and anticipate publication in a peer-reviewed journal upon completion.

THE "MSKCC" STUDY

The third API epidemiology investigation to be presented, the "MSKCC" study, is well characterized by its complete title: "A Prospective Mortality, Morbidity and Tumor Registry Study in Refinery, Petrochemical, Laboratory, and Research and Development (R&D) Workers". The rationale and methodology of the study were presented by Dr. David Schottenfeld in London at the Royal Society of Medicine at the 1980 Anglo American Conference on Human Health and Environmental Toxicants (4). Dr. Schottenfeld commented that the study was "unique in its scope and dimensions", and noted that, "it provides an exciting challenge for assessing the health status of employees within selected segments of a major national industry."

Initiated in 1977 in three plants in order to establish feasibility of methods and procedures, the study expanded to 20 companies with 110 plants and facilities and included some 120,000 employees and annuitants in its register.

As a prospective study there was, of course, no backlog of information for entry into the system, and data were incorporated as the events occurred. A preliminary analysis of mortality surveillance was carried out, based upon the 502 deaths accumulated during the period 1977-1980 (122,607 man-years of observation)(5). As is true of prospective studies during the early reporting periods, standard mortality rates tend to be lower than those of the U.S. male population for all major causes of death, including heart disease, strokes and accidents. The preliminary results of this study were consistent with that pattern. Mortality due to cancer as a whole, and lung cancer in particular, was also lower than the expected norm. It must be emphasized, however, that any consideration of these results must reflect the highly tentative and preliminary nature of the analysis. More definitive results and inferences would require a longer period of observation and a thorough validation of the data base.

Having completed five years of initial operation of the "MSKCC" study at the end of 1981, a re-assessment of the present status and future direction of the research is now taking place. It is recognized that a surveillance system with entry data dependent upon voluntary participation by the companies results in a study which is vulnerable to questions and challenges. Consequently, we plan to transform the surveillance type of operation into a validated prospective study of a defined population. Reporting during the validation stage and during the foreseeable future will be limited to mortality experience and census data in refinery workers only, with the investigator documenting the data. This will reduce the scope to more manageable proportions and results in a more credible study.

CANCER AS A CAUSE OF MORTALITY IN REFINERY WORKERS

During the latter part of the decade of the 1970's, occupational and environmental cancer achieved a remarkable degree of prominence (perhaps notoriety is more descriptive!) in the eyes of labor, interest groups and the general public. With crude petroleum and certain of its derivatives known to contain established carcinogens and to exhibit positive results in carcinogenesis bioassays, concern about occupational cancer in refineries is understandable. In this setting, it was inevitable that epidemiologic techniques would be applied to refinery worker populations to seek out any increase in cancer occurrence. Indeed, the determination of cancer incidence in refinery workers has been the major justification for the API epidemiologic studies, for the IP epidemiology research carried out by Professor Alderson, for the Australian Institute of Petroleum's Healthwatch project, and for a number of investigations conducted by governmental and academic epidemiologists.

Certain epidemiologic investiations of occupational cancer can be classed as "Hypothesis-Generating", and others as "Hypothesis-Testing". Two "Hypothesis-Generating" reports developed by epidemiologists of the U.S. National Cancer Institute attracted a great deal of media attention. Both attributed a higher cancer incidence to refinery work. The first of these, released by Blot et al in 1977, reported cancer mortality in U.S. counties having petroleum industries, in comparison with counties without petrochemical operations (15). Blot reported excess cancers principally of lung, nasal cavity and skin (including melanoma). (The study failed to take note of the fact that many of the "petroleum" counties also contained shipyard facilities with associated exposures to asbestos, a notorious causal factor in cancer of the lung and other sites.) In the second "Hypothesis-Generating" study, Thomas et al released reports in 1980 and 1982 on cancer mortality in refinery workers based upon death records collected by the Oil, Chemical and Atomic Workers Union (OCAW)(16) (17). Higher cancer incidence was reported for stomach, kidney, brain, leukemia and multiple myeloma, with suggestive increases for liver, pancreas, lung and skin. These studies have been criticized for a variety of methodological reasons, principally the use of Proportionate Mortality Ratios (PMR's) which are less definitive than the Standardized Mortality Ratios (SMR's) for which the study population is carefully established. Unfortunately the fact that the intended purpose of these studies was "Hypothesis-Generating" tended to be overlooked in news media reports, which often portrayed the findings as gospel.

To move on to "Hypothesis-Testing" studies, I wish to summarize six reports on cancer as a cause of death in refinery workers. Two are studies of Canadian refineries, three from the U.S. (including the "Tabershaw" study), and the sixth is the IP study of refineries in the U.K. by Professor Alderson.

The studies, listed in chronological order, are identified and their positive finding summarized as follows:

I   Hanis, et al (1979)(18)(19)   : Imperial Oil Limited
                                         21,737 Male Employees or
                                         Annuitants
                                         1964-1973

        Significant Cancer Increase   : Lung* and Gastrointestinal
                                         Tract**

II   Theriault and Goulet (20)   : Shell Oil Canada
        (1979)   1,205 Male Employed 5 years
                                         1928-1976

        Significant Cancer Increase   : Brain (3 deaths)

III   Rushton and Alderson (13)(14) : 8 Refineries in Britain
        (1980)   34,781 Workers Employed at
                                         least 1 year
                                         1950-1975
                                         4,406 Deaths; 99+% Follow-up

        Significant Cancer Increase   : Nasal Cavity and Sinus
                                         (O=7; E=3.1) and Melanoma
                                         (O=14; E=6.5)

*   In petroleum workers exposed to crude oil and petroleum products
    in comparison with workers without such exposures.

** In refinery workers in comparison with non-refinery workers.
    However, when workers are compared with the Canadian male
    population, there are no significant increases in cancer as a
    cause of death.

IV   Tabershaw/Kaplan (1982)   : 17 U.S. Refineries
                                         20,131 Male Workers
                                         1962-1976
                                         2,237 Deaths; 97% Follow-up

        Significant Cancer Increase   : "Cancer of Other Lymphatic
                                         Tissue"

V   Hanis, et al (1982)   : Exxon Refinery, Baton Rouge,
                                         Louisiana
                                         8,666 Workers and Retirees
                                         1970-1977
                                         1,199 Deaths; 91.4% Follow-
                                         up

        No Statistically Significant Increase in Site-Specific
        Cancer Deaths

VI   Wen, et al (1982)   : Gulf Oil Refinery, Port
                                         Arthur, Texas
                                         16,880 Workers
                                         1937-1978
                                         4,358 Deaths; 94% Follow-up

        Significant Cancer Increase   : Bone (9 Deaths)

Figure one presents in summary fashion the findings of the
"Hypothesis-Generating" and "Hypothesis-Testing" studies. I would
emphasize that, in a strict scientific sense, the studies are not
methodologically comparable. Differences in protocol, time and
duration of observation period, geographic location, size and ethnic
make-up of population, selection of controls, statistical methods
used in analysis and other factors preclude numerical comparison of
results. The figure indicates the types of cancer, reported by the
authors to cause death in refinery workers, which occurred at a
higher incidence than expected with a 95% confidence level. Certain
of the positive findings reported in the "Hypothesis-Generating"
and "Hypothesis-Testing" studies need to be investigated more fully
by additional and/or expanded studies, case control follow-up, etc.
In this simplistic presentation of positive correlations, I have
ignored significant negative correlations in certain studies, as
well as the magnitude of the association and the power of analyses.
Even so, there is striking lack of concurrence in these multiple
studies in identifying site-specific cancer excesses as a cause of
mortality in refinery workers. This argues strongly against the
presence of a serious and consistent occupational cancer problem
in petroleum refineries.

NEW API EPIDEMIOLOGY STUDIES

While continuing the retrospective and prospective mortality studies
as previously mentioned, the API is currently initiating
epidemiology studies in two new areas, relating to, first, the
so-called "Healthy Worker Effect", and second, methodologies
applicable to assessing the possible effects of occupational
factors upon reproduction.

THE "HEALTHY WORKER EFFECT"

Mortality studies in the petroleum industry, as in most occupational
studies, tend to show quite favorable experience with fewer
observed than expected deaths in many diagnostic categories. While
a number of explanations are offered to explain this "healthy
worker effect", it has not been evaluated in a thorough scientific
manner. Dr. Brian McMahon and Dr. Richard Monson at Harvard are
now undertaking a critical review and analysis of the subject, and
API is providing certain data bases which will enable the
investigators to link their studies to actual experience in the
petroleum industry.

METHODS FOR EVALUATING REPRODUCTIVE EFFECTS IN WORKERS

There is growing concern about occupational exposures in both men
and women having possible adverse effects upon reproduction,
particularly pregnancy outcome. This stems in part from the
teratogenic effect of certain drugs and chemicals in human subjects.
While most petroleum hydrocarbons which have been studied were not
teratogenic in rodents, it may be necessary or desirable to monitor
worker populations for possible reproductive effects. Such studies
would be very complex and difficult to carry out. API is supporting

Dr. J. Kline of Columbia University in conducting a comprehensive evaluation of methodologies and data bases which might be applicable to studies of possible reproductive effects in worker populations.

U.S. OCCUPATIONAL EPIDEMIOLOGY IN PERSPECTIVE

Of the various disciplines of occupational medicine - which include toxicology, industrial hygiene and clinical studies, for example - epidemiology is best suited to detect the subtle, chronic health effects in employees which may result from exposures in the work environment. In recognition of its high potential value, occupational epidemiology has grown apace in the U.S. during recent years. As recently as 7 or 8 years ago, there were no full-time practitioners of the discipline in the petroleum industry. Today I can count more than twenty-five epidemiologists employed under member companies, and a little over a year ago I was pleased to add an epidemiologist to staff at API.

In this period of rapid growth, occupational epidemiology in general has yet to reach maturity. I would liken the present state-of-the-art in the U.S. to a gangling adolescent very much in the awkward stage - full of dreams, hopes and aspirations; often doomed to results which fall short of the mark if not outright failure, but with some resounding successes. Investigations in occupational epidemiology usually result in a spectrum of results, ranging from definitive positive to negative associations, but with a region of uncertainty in between. This uncertainty leads to ambivalence and ambiguities in interpretation which are troublesome. Most occupational epidemiology studies are retrospective in nature and face difficult, if not insurmountable, problems of gaps and difficulties in the information available, particularly in reconstructing exposure histories. The usual response is to use job assignment as a surrogate for exposure, but this often poses difficulties of equal, if not greater, severity. Fortunately, industrial hygiene measurements of the working environment are now increasingly available and the ability to accumulate industrial hygiene data and other relevant information in a planned, rational manner represents a decided advantage in prospective studies. However, these studies may require long periods of time - a decade or two - to achieve meaningful results.

Since the worth of epidemiologic investigations is highly dependent upon the completeness and validity of records available for analysis, the new generation of medical record systems holds great promise. Utilizing recent advances in information management, especially the prodigious memory capacity of the newer electronic data processing equipment, it is now feasible to collect, store, integrate and recall for analysis the clinical medical records, industrial hygiene measurements, life-style and other pertinent information accumulated throughout the working history of each employee.

Perhaps the advanced medical records system will provide what is needed for occupational epidemiology to come of age and reach the expected goals.

CANCER AS A CAUSE OF DEATH IN REFINERY WORKERS
(SUMMARY)

Cancer Site	Lung	Naso-Sinus	Skin (Incl. Melanoma)	Esophagus	Stomach	Intestines (Incl. Rectum)	Liver	Pancreas	Kidney	Brain	Leukemia	Myeloma "Other Lymphatic"	Bone
Hypothesis Generating (2)	=	=		—		—	—	—	—	—	—		
Hypothesis Testing (6)	—	—	—	—	—				—		—	—	

54

REFERENCES

1. An Epidemiological Study of Cancer Among Employees in the American Petroleum Industry. Final Report. The Kettering Laboratory. API Med. Res. Pub. (1958).

2. A Mortality Study of Petroleum Refinery Workers: Social Security Follow-up. Final Report. Tabershaw/Cooper Associates, Inc. API Med. Res. Pub. (1975).

3. Mortality Study of Petroleum Refinery Workers: Update of Follow-up. Draft Final Report. S.D. Kaplan, SRI International. (February, 1982).

4. Schottenfeld, D. et al. Epidemiological Surveillance of Petroleum Refinery Workers, Human Health and Environmental Toxicants: Royal Society of Medicine International Congress and Symposium, Series No.17, published jointly by Academic Press, Inc. (London), Ltd., and the Royal Society of Medicine.

5. Schottenfeld, D. et al. A Prospective Study of Morbidity and Mortality in Petroleum Industry Employees in the United States - A Preliminary Report. Proceedings of the Banbury Conference on Qualification of Occupational Cancer. Cold Spring Harbor Laboratory (1981).

6. Leitsch, A. Paraffin Cancer and its Experimental Production. Brit. med.J., 2. 1104-1106 (1922)

7. Henry, S.A. Occupational Cutaneous Cancer Attributable to Certain Chemicals in Industry. Brit. med. Bull., 4, 389-401 (1947).

8. Twort, C.C. and Twort, J.M. The Carcinogenic Potency of Mineral Oils. J. Indust. Hyg., 13, 204-226 (1931).

9. Shapiro, D.D. and Getmanets, I.Y. (Blastomogenic Properties of Different Sources). Gigiena i Sanitoriya, 27, 6, 38-40 (1962).

10. Horton, A.W. et al. Composition versus Carcinogenicity of Distillate Oils. American Chemical Society, Division of Petroleum Chemistry Preprints, 8, No.4C, 59-65 (1963).

11. Bingham, E., et al. The Carcinogenic Potency of Certain Oils. Arch. Environ. Health, 10, 449-451 (1965).

12. Medical Research Council Great Britain. The Carcinogenic Action of Mineral Oils: A Chemical and Biological Study, MRC Special Report Series No.306 (1968), London, HMSO, pp251.

13. Rushton, L. and Alderson, M.R. An Epidemiological Survey of Eight Oil Refineries in the U.K. - Final Report. Institute of Cancer Research, Surrey, England 28.3.80.

14. Rushton, L. and Alderson, M.R. An Epidemiological Survey of Eight Oil Refineries in Britain. Brit. J. Ind. Med., 38, 225-234 (1981).

15. Blot, W.J., et al.   Cancer Mortality in U.S. Counties with Petroleum Industries. Science, 198, 51-52, (1977).

16. Thomas, T.L. et al.   Mortality Among Workers Employed in Petroleum Refining and Petrochemical Plants. J.O.M., 22, No.2, 97-103 (1980).

17. Thomas, T.L. et al.   Mortality Patterns Among Workers in Three Texas Oil Refineries. J.O.M., 24, No.2, 135-141 (1982).

18. Hanis, N.M. et al.   Cancer Mortality in Oil Refinery Workers. J.O.M., 21, No.3, 167-174 (1979).

19. Hanis, N.M.   A Study of Intestinal Cancer Mortality in a Population of Canadian Oil Workers. Thesis, (1977).

20. Theriault, G. and Goulet, L.   A Mortality Study of Oil Refinery Workers. J.O.M., 21, No.5, (1979).

21. Hanis, N.M. et al.   Epidemiologic Study of Refinery and Chemical Plant Workers. J.O.M., 24, No.3, (1982).

22. Wen, C.P. et al.   Long-Term Mortality Study of Oil Refinery Workers. I. Mortality of Hourly and Salaried Workers. Submitted for publication (1982).

OPENING ADDRESS

Dr. K.P. Duncan
Director of Medical Services,
Health and Safety Executive.

INTRODUCTION

In two years' time the Health and Safety at Work Act will be ten
years old. Some six years ago the Institute of Petroleum launched
a conference on the theme of the Act. They were not alone in doing
this and their action reflected the expectation of many that something quite fundamental had changed in the field of health and
safety and that there were new developments which required careful
assessment and monitoring. Different people expected different
things from the Act and it is worthwhile considering briefly what
the various expectations were and how they have been realised.
Perhaps then we can consider what are the likely developments over
the next eight to ten years. It is encouraging to see a programme
which covers so much work which is already going on in the industry
and covers such a wide spectrum of types of activity. How far that
in itself relates back to the coming into force of the Act would be
hard to tell but undoubtedly it did act as some kind of stimulus
and equally undoubtedly many of us were aware that if the opportunity afforded by this new incentive was not taken a great chance
would be lost. How far has it operated as expected, how far have
we succeeded?

EXPECTATIONS IN THE 1970'S

It is possible to describe quite a number of headings for these
expectations but for our present purposes four will suffice.

(a) There would be a change in the legislative pattern which
     would, to some extent at any rate, follow the recommendations
     in the Robens Report.

(b) In particular there would be an attempt to get more worker
     participation and involvement in the field of health and
     safety.

(c) There would be a need for much more research and deeper study
     which would link, at least in part, to the development of new
     technology though it should not ignore the continuation of
     older hazards.

(d)   The various roles of different health professionals (doctors,
      scientists, nurses, epidemiologists, hygienists and so on)
      would probably change.

Let us examine these four headings separately, considering what
the expectation was and briefly how it has been realised.

CHANGE IN THE LEGISLATIVE PATTERN

The 1974 Act had a considerable number of new features and, if one
combines with it the Safety Representatives Regulations made
consequent on the Act, these represented quite fundamental changes.

Participation and consultation can be jargon words of little
significance. However, if this degree of involvement showing
actual concern in involving those accepting risks, was to be successful, the Act and the organisational consequences of it provided an
excellent framework. The tripartite nature of the Health and Safety
Commission and the way in which that was to operate was obviously
central to the success of the underlying philosophical objectives
embodied in the Act. This can probably be best illustrated by
consideration of the standard setting process. The underlying
change in our approach to toxicity, and perhaps particularly
carcinogenicity, and the growth of the no-threshold assumption,
meant that there had to be a very different attitude in the setting
of limits of control. What actually happens in practice is that
when a substance is being considered as part of the standard setting
process the first thing to do is to get a very careful review of the
present scientific position. This review has to cover all sorts of
basic data and it is a sad fact of life that these data are always
incomplete. Furthermore many of them are also difficult to interpret because of biological variability between species or individuals
or other factors and there is nearly always a shortage of good human
information. However, the best attempt has to be made to get the
science right and to record that. That process is a purely scientific and expert one.

At the same time as that is going on a study has to be made of the
most effective methods of control that are operating in industry
and that is also essentially a factual matter. The third piece of
information which is needed for the standard setting process is
some consideration of the economic and social aspects of a problem.
It is, therefore, not difficult to see that a tripartite body made
up of employers, employees and government is a perfectly sensible
and logical one to approach this amalgam of different kinds of
knowledge. It is equally obvious that the role of the expert has
to be clearly seen in these developments. The expert is very
important because if the science is not right then whatever is
concluded will be wrong, but equally the expert is only one citizen
when the decisions which are taken move out of his expert field.
This has been very well reviewed by John Locke in a recent Redgrave
Lecture. As an aside, this does make life both more interesting
and more difficult for the expert but it does give reasonable
chance of involvement of those who are to be exposed to the risk

rather than any assumption that they can leave these decisions to
somebody else.

This means that there has to be a great deal of openness and spread-
ing of information and a great progressive educational development
among those involved in the process. It is an interesting consequence of this need that a number of larger trade unions have now
themselves appointed health and safety expert advisers to interpret
these matters to their colleagues.

The other big change in the legislative approach has been the
development of the Notice procedures. It had been acknowledged for
a very long time that the penalties under prosecution were rather
out of proportion to the resources of those paying them or perhaps
in many cases to the severity of the event. The introduction of
Improvement Prohibition Notices has undoubtedly put a very powerful
tool in the hands of the inspectorates and the statistics that show
that very few of these notices have been sucessfully appealed
against would I hope be able to be intrepreted as an indication
that the powers have been sensibly and correctly used.

The third legislative change that I would draw attention to is the
concept of a mixture of regulations, approved codes and notes of
guidance, a hierarchy of documents with slightly different purposes.
As a personal opinion I would think that the involvement of more
people in the standard setting process has worked well but
I would not be quite so sanguine about the understanding of the
differences between these three categories of document. The hope
would be that regulations could be very brief, rather general, that
the approved codes could be written in much more flexible style
with recommended methods of operation which were not themselves
mandatory provided it could be shown that some alternative method
was as effective. To some extent this has not moved as smoothly
as would have been hoped. On some topics, such as guidance on
VDUs, the degree of anxiety about drafting niceties has been dis-
proportonate to the extent of real hazard. The reason is probably
two-fold. One the one hand it is surprising how many employers
and indeed the professionals working in the health and safety
field, want things spelt out in far more detail than some of us
would think right. On the other hand, documents written to be
helpful, because they come perhaps from a government source, are
sometimes looked at with a degree of scrutiny which is really only
applicable to the strict requirements of a regulation. These two
features have made it much harder to do some of the drafting and
some of the putting into effect than perhaps one would have anti-
cipated. It may be that this process must take time and that it
is wrong to be a little impatient about some of the delays.
Certainly we would be foolish not to try to keep this whole process
working by consensus.

MORE WORKER PARTICIPATION

We have considered briefly the question of worker participation
and involvement in the standard setting process and in the determination of appropriate control limits. The full consequences of that
are outside our present consideration. At the level of the factory
or other place of employment there has been in many cases a very
successful development of the use of safety representatives and of
safety committees. The trade unions ran a very extensive training
programme for these representatives and while obviously anything on
that scale is bound to be a little patchy, there has been a great
spread of information and an improvement in understanding. Equally
I think there has been an appreciation by employers of the importance of conducting their studies into health and safety problems
in the open. All of this is entirely laudable but it does lead to
very obvious difficulties when information is partial, uncertain
and where it can be used to press what are essentially industrial
relations points rather than health and safety considerations. We
have not got as far as we doubtless have to go in getting an
appreciation of fairminded objective analysis of potentially hazardous situations which can be accepted by both sides of industry.
There is a very real danger that health and safety services provided
by industry itself will come under the threat of misunderstanding
unless they conduct themselves with scrupulous care and openness.

The word "independence" is often very loosely used. It is alleged
that medical and scientific services provided inside industry,
because they are paid for by that industry, cannot be independent.
This is a very central argument and personally I feel it insulting
to suggest that because someone is employed in one particular place
he will automatically distort his scientific appreciation of a
situation. But let nobody be under any illusions, this is an
accusation which is being made. Probably it is fair to say that
nobody is really independent, even academic departments, because
even they depend on attracting the research funds or grants. I see
no easy solution to this. It is quite obvious that organisations
like my own will never have the resources to do all the work that
is required and it is equally sensible that industry should be
contributing to these studies. I think only by constantly
demonstrating scientific integrity can this obstacle be overcome,
but obstacle it is. Employees must be involved early in the
selection of topics for study and in the development of epidemiological protocols and research projects.

RESEARCH

Research of all kinds, including epidemiology, has increased in the
occupational health field a great deal over the last eight years.
This meeting itself, and its programme, bears witness to that and
some of the large scale epidemiological studies, such as those
reported here, would probably not have taken place but for the
stimulus of the coming into being of the Act. I suspect that
statement can, and probably will be challenged but I think it is defens-
ible. The period under review has seen also the controversy over

the Rothschild provisions and the development of the customer contractor principle. From the point of view of the Health and Safety Executive we have found this works well with our main contractor, the Medical Research Council; while possibly it could be argued that it has introduced a little more administration into the system that is I think now minimal. The gain in direct access more than cancels that and the fear that only applied research would prosper under such a system has certainly not proved to be so if you look at our own research programme over the last five or six years. As mentioned earlier, it is not only the new technologies that have acted as a stimulus but new developments in laboratory techniques have opened up whole new approaches to the study of toxicology and carcinogenicity. There is still room for much more work. I think it needs to be discussed how much we need to direct some of our epidemiological effort. Clear results can be expected where the disease concerned is rare or where the incidence is very high indeed but that is not often the case.

It also needs to be discussed how we present results of marginal significance or marginal apparent significance to the lay man, to our audience of workers. There is a feeling often expressed that provided we could keep far more detailed records of everything to which everybody was exposed and put them all on to computers we will in reasonably brief time be able to comprehend far more about the interaction of man and his working environment than we can at the moment. This is a very doubtful proposition. What we are going to have to handle are a lot of doubtful associations, which may not be real, may be part of a compound series of events and we have to try to explain these in ways that bring caution without excess anxiety or relaxation without licence.

It is very very difficult to loosen controls once they have been imposed and it is always easier to ratchet things tighter rather than slacker. In many circumstances this may not matter very much but there are cases in which progress could be unnecessarily hindered by over-anxiety. This is a view constantly put forward to us by the scientific community but I think it is fair to say that when challenged to produce examples where this has caused obstruction to progress the scientific community have not been very successful. Undoubtedly, however, there are occasions where the considerations of health and safety have increased costs but surely it is not going to be argued that there are quite a lot of occasions when that should be true. What one would hope from more research and more understanding would be that there would be a better perception of risk on a very widespread scale and that therefore the whole business, not only of setting the standards, but of accepting them would become much easier. That has, of course, to be a long-term thing and the psychological problems which interfere with that degree of objectivity are very considerable. Perhaps the most conspicuous could be described as "guilt by association" where substances have been connected with war or other disastrous events attain a reputation perhaps beyond that that they would earn from a strictly scientific point of view. The difference between measured and perceived risk is very real.

CHANGING ROLES OF HEALTH PROFESSIONALS

Much of occupational health practice stems from working conditions
typified by a picture of industrial Britain in the 19th Century.
There was concern with child labour, with the need to protect
certain classes of society and with the need to control and treat
occupational disease. There has been a revolution in environmental
conditions in the last twenty or thirty years. There are still some
fairly black spots and some pretty horrible bits of industry but by
and large people do not work in anything like as unpleasant conditions as they did a generation ago and the incidence of significant
frank industrial disease has dropped. At the same time, of course,
there has become increasing concern with the longer latent period
of more insidious disease. It seems to me at any rate that with
these developments the role of the doctor has to be looked at
again. He does not have an automatic entitlement to be considered
as the leader of the occupational health team in all circumstances.
Certainly he does not have any automatic right. Some of our
thinking has been much influenced by the developments that happened
after the Windscale incident in the United Kingdom Atomic Energy
Authority. After the Fleck Report the developments in the Authority
were along the line of setting up integrated health and safety
organisations which included doctors, scientists, engineers, admin-
istrators, biologists and physicists. In one place the doctor
might be head of the team, in another it might be the physicist,
and this approach of using a multidisciplinary group seems to be
essential in technical industries. At the same time, when frank
industrial disease is much less the object of the exercise, the
need for many of the classical routine medical examinations disap-
pears. These changes have not been easily accepted but they are
gradually developing and the emphasis on environmental measurement
has increased. However, this can go too far. It is very important
that the relationship of environmental measurement to actual intake
and the relation of that intake to biological effect and the relation of that effect to long-term damage has to be most carefully
worked out. This is perhaps one of the fields for a great deal more
study in the future.

WHAT ARE THE LIKELY FUTURE DEVELOPMENTS?

I hope that what I have said has sounded fairly optimistic because
basically it has not been a bad six years since you last had a
conference. However, we cannot ignore the influence that the
European Community will be bound to have on our own legislation and
it would be pointless to deny that there are differences in approach
between the one we have traditionally accepted and that which is
more common in the other nine countries, or at least in some of
them. We have tended to go down a pragmatic, rather practical line;
our legislation contains phrases that are hallowed in interpretation
in English law such as 'reasonably practicable'; we are operating
or attempting to operate very much in this system of consensus
decision. Perhaps we have gone further along the road of arguing
with experts than is common in some countries. So there is
obviously a great deal of dialogue to go on there and it does go on.

There are certain other aspects, technical, governmental or legislative ones, which have changed. I have mentioned the interdisciplinary point. There is a need for more recognised and clearer defined qualifications, not only in medicine and nursing, where considerable steps have been taken, but in occupational hygiene, in safety and in other fields. There is a need to strengthen the academic base on which all these things depend: it was sheer serendipity that brought most of us of my generation into this particular activity. We were mostly mid-career shifters, and it is necessary now with a more scientific approach, that people will come, perhaps in a more orderly way to accept that there are changing managerial and other demands. Then again we have to do so much more in the open and we have to do so much more under criticism, possibly good for us, than we had to twenty or so years ago. That is all changed. We have to face the arguing out of some of these difficult decisions in public. That is not difficult at the level of an industry, you can talk to the people who are actually involved. It is very difficult when you are dealing with queries at a more political level, either with union leaders who are no longer actually involved at that particular work place or with an organised management representative who does not have quite such close contact with his workforce. All this is dominated by the over-riding influence of the much more coercive media, to which we are now subjected. It is necessary for us all to understand exactly what the problems of the communication of advanced science are to millions of people who know very little about it. I have great confidence in the capacity of health and safety

experts at work to explain to that particular workforce, who know quite a lot of what is going on; never underestimate what they do know. I have much less confidence in the capacity to explain these problems essentially in the abstract and nationally.

So we are left with the situation that we have had a fairly fundamental change, we have cashed in on it fairly successfully in extended studies, in extended involvement in health and safety. We have developed a bit along interdisciplinary lines and we have accepted, not totally, but very largely, the need for much more public pronouncement about what we are doing and the need to do our studies in the open. It will be interesting, if in six or more years time you have another of these conferences to see how far these prophecies are proved right and how far the path, which has been quite a successful one, has been continued.

INSTITUTE OF PETROLEUM HEALTH RESEARCH
ACTIVITIES

Dr. W.L.B. Leese

Corporate Medical Adviser, Britoil plc

INTRODUCTION

The Institute of Petroleum has now been associated with occupational
health research for forty years. The common theme running through
much of this activity has been concerned with the carcinogenicity of
certain mineral oil fractions and the derivation of appropriate
safeguards.

Mineral oil was added to the list of acknowledged carcinogens in
1914. In Britain, experimental skin cancer was produced in animals
with shale oil in 1922, and later that decade the carcinogenic
action of mineral oil was demonstrated also in animals. It was
during this period that the Manchester Committee of Cancer was
set up to examine the cause of mulespinners' cancer in cotton mills.
From 1943-1946 the then Petroleum Board and the Institute of
Petroleum supported investigations at the University of Birmingham
which showed that solvent extracts and other fractions of mineral
oils were highly carcinogenic to mouse skins. The work of
Cruickshank and Squire at Birmingham and Manchester Committee's
assessment, based on the experience of the Petroleum Board, the
Institute of Petroleum and individual oil companies, led to the
drawing up of a Factory Act specification on highly refined white
oils. This was introduced for spindle lubricants in 1953, and
replaced the earlier Twort specifications of 1945.

Although health research has been supported by the Institute of
Petroleum since this early period, its first in-house investigation
was carried out by an Ad Hoc Committee set up in 1967 to review the
effects of mineral oil on health. This Committee was working in
parallel with a Medical Research Council study which produced a
Report drawing attention to the carcinogenic action of certain
mineral oils.* The Institute of Petroleum's Ad Hoc Committee
published its own recommendations in 1969 as a document called
"Mineral Oil and Skin Cancer".** This recommended the use of
solvent-refined oils or oils which have received a severe refining

*   MRC Report - SRS 306. HMSO, London 1968.
** Petroleum Review, 1969, 23, 311-12.

treatment to reduce aromatic content. These were considered to be less hazardous than unrefined oils. By using such refining methods, and provided sensible hygiene precautions were adopted, the risk of skin cancer arising after prolonged and repeated use of mineral oils was minimal.

The Institute of Petroleum's Advisory Committee on Health developed from this Working Group in 1969. It is a multi-disciplinary committee of oil industry Medical Officers and a consumer group doctor, industrial hygienists and toxicologists, technical service specialists and representation from the marketing side of the industry. It has four specialist sub-committees of medicine; industrial hygiene; toxicology and a sub-committee of analysts.

It could be asked why in addition to individual oil companies the Institute of Petroleum got involved in research work. Many of the companies in the lubricating oil business are small and do not have the resources of expertise or of funds for extensive research. The Institute of Petroleum exists to give objective consideration to the science and technology of petroleum, its products and their uses. It is more rational to carry out this kind of work on a centralised basis where knowledge can be pooled and advice offered in a detached and overall form.

When the Institute of Petroleum decides on a specialist investigation, it is the practice wherever possible to fund independent consultants who are acknowledged experts in the field. The protocol is agreed, they carry out the work and report to the Institute, but have always had the freedom of scientific publication.

During this whole period of involvement with the health aspect of mineral oils, the factors which have stood out are:-

1) The refinement of analytical techniques – in particular, the development and use of gas liquid chromatography in the identification and quantification of polycyclic hydrocarbons.

2) The continued need to fall back on animal testing, even although this does not replicate the conditions of human exposure.

3) The similarity of advice after each piece of research on mineral oils:-
    (a) Minimal contact.
    (b) Good personal hygiene.
    (c) Good plant hygiene.
    (d) No misuse of product.

But these are no more than standards of good occupational health.

THE BIRMINGHAM UNIVERSITY REPORT OF A STUDY OF OCCUPATIONAL SKIN CANCER 1975

One of the early research projects which the Advisory Committee on Health recommended the Institute to support financially was that at Birmingham University by Brown, Waldron and Waterhouse which investigated the occupational histories of 298 patients registered at the Birmingham Cancer Registry between 1936 and 1972 with a diagnosis of scrotal cancer. These scrotal cancer cases were 1.7% of all skin cancers. Skin cancers were 11.4% of all tumours registered. In a population of 2.5 million males, this represents an incidence of scrotal cancer of c. 5 cases per million males per year. This was about three times the national average, but nevertheless a rare disease.

Occupational histories were obtainable for 109 of the 298 patients (mainly for those registered since 1952). Ninety-four of the one hundred and nine had an exposure to oil, of whom forty-two were toolsetters or machine operators. The working exposures relate back many years when the origin and refinery methods used in the preparation of oils are uncertain, and for instance may have included oils of shale oil derivation. Neither was it possible to determine to what extent hygiene practices may have been observed at such an early period.

Among these 298 scrotal cancer patients, 52 developed other primary tumours, 10 before and 42 subsequent to the scrotal cancer. Eleven of the twelve patients with subsequent bronchogenic carcinoma, had an oil related occupation. It was not possible retrospectively to assess the relevance of smoking as a causative factor in these cases, but it was suggested that there might be in some individuals a genetic cause for an increased susceptibility to hydrocarbon carcinogens. In this connection, it is of note that in the Epidemiological Surveys described by Dr. Alderson on Oil Refinery and Distribution Centre employees together numbering over 58,000 men who were followed during a period of 25 years, there were no scrotal cancer deaths and a deficit in deaths from epithelioma, and there were an observed 800 deaths from lung cancer against an expected 1015.53.

LEAD IN AMBIENT AIR

In the early 1970's the Institute of Petroleum in co-sponsorship
with government departments, helped to support financially the U.K.
Atomic Energy Authority Harwell Investigations into human lead
up-take from motor vehicles' emissions. This Report was published
in 1978.*

The object was to estimate the up-take by humans of lead from petrol-
driven vehicles and the contribution by this to the total up-take
of lead. Radioactive tracers were used to determine the effect of
chemical composition and particle size of lead aerosols on up-take,
and also to compare up-take from lung and gut and subsequent
excretion routes and rates. The effect of lead on health was not
included in the study.

The conclusions indicated that the contribution of air lead to
total up-take at the highest levels to which people are
continuously exposed (for example, in houses adjacent to the busiest
stretches of a motorway) would appear to be comparable with that
from diet (about 16%).

However, for the majority of urban and rural residents, for whom
airborne lead averages less than 0.5 μ/m^3, the data indicated a
contribution of air lead to total up-take of less than 10%.

This lead study was one of very few where the results have been
based on the monitoring of the up-take by human volunteers who
deliberately exposed themselves to lead and to radioactive lead
aerosols of known but ethically acceptable concentrations.

The Advisory Committee on Health has been reviewing the substantial
literature on the health effects of lead, which have been published
up to and since the Lawther Report of 1980. Much of this has been
directed to total environmental lead and child behaviour.

At a recent Symposium of CLEAR (The Campaign for Lead Free Air), a
group set up with the declared objective of securing the removal of
lead from petrol, Dr. Rutter, a Professor of Child Psychiatry, (who
was on the Lawther Committee), expressed his personal views that
there was - "....a probability, not a certainty, that low lead level
exposure may have adverse psychological effects. These are not
large and there are clearly many other important influences
involved. It is likely that there are considerable differences in
individual exposure related to age, ethnicity, social variables and
factors as yet not understood." He went on to say that in his view
this justified the removal of lead from petrol, but then stated,
"....removal of lead from petrol will not bring about a major
improvement in the health and development of our children and if we
claim that, we raise false expectations and run the real danger of
a backlash when those hopes are unfulfilled."

* "Investigation into lead from motor vehicles". Chamberlain et al.
  HMSO.

The Advisory Committee on Health does not share Professor Rutter's view that the present evidence is sufficient to justify action beyond the present declared policy of the U.K. Government, namely to reduce the maximum lead level in petrol to 0.15 g/litre at the end of 1985. This should, of course, reduce the lead from petrol in the environment by virtually two-thirds.

A key question to be answered is whether the stated adverse behavioural effects reported in association with the relatively high blood lead levels found in U.S. children in the past, occur at the much lower levels which are typically found in British children today. Two further M.R.C. - supported U.K. studies on this subject are due for publication this autumn and will merit careful scrutiny.

The Institute would welcome an authoritative assessment by the Royal Commission of the contribution made to airborne lead, by lead in petrol. Does it contribute 10%, 20% or more to average body intake? In addition, appropriate surveys of air and blood levels both before and after the planned reduction of lead in petrol to 0.15 g/litre would prove of value to obtain objective evidence of the change. Thereafter, it would be easier to assess where further reductions of environmental lead would prove helpful, and where priority of action should lie.

THE EPIDEMIOLOGICAL SURVEY IN THE OIL INDUSTRY

The Epidemiological Survey by Alderson and Rushton of the eight U.K.
Oil Refineries (1980), The Distribution Centres Study (1982) and
the London Transport Maintenance Department Study (1982) constituted
the largest single health research project undertaken through the
Institute of Petroleum. As Chairman of the Survey's Steering
Committee, I wish to take this opportunity to thank the twenty-three
oil companies who acted as sponsors of the project; who, for six
years, gave increasing amounts of money to meet inflation to see the
study to completion. I wish also to pay tribute to Dr. Alderson and
Dr. Rushton for their steadfast scientific dedication over so many
years.

The success of any 'a priori' study must depend on the quality of the
data available for analysis. Given time and patience, mortality
surveys, as here, can be very complete. The attempt to categorise
employees according to their work exposure details is more difficult
as personnel records in the past have so often been maintained
largely for the purpose of performance, recompense and promotion.

It is, therefore, relevant to draw attention to a health surveillance
system being established by the Australian Institute of Petroleum
to cover 12,000 refinery and terminal workers. This is a prospective
study which is to include personal interviews with each subject
to provide a precise job description, together with information on
confounding variables which include smoking and drinking behaviour.

The objectives are to cover mortality and cancer morbidity with
statistics and data collection to continue on employees after they
leave the industry.

INSTITUTE OF PETROLEUM'S THREE-YEAR HEALTH RESEARCH PROGRAMME

The Advisory Committee on Health launched its other main Health Research Programme in 1978. There were five separate projects which are just now completed. I will refer to three of these.

Analytical Work on Used Cutting Oil and Oil Mists. Following the suggestion that polycyclic aromatic hydrocarbon content of metal working oils may increase during use and may thereby present greater potential carcinogenicity from skin contact and mist inhalation, one project was to investigate the potential increase with use in cutting oils, quenching oils and oil mist arising from their use.

There are a large number of different polycyclic aromatic hydrocarbons, and something less than twenty of these have so far been found to be carcinogenic in skin painting tests in animals, (benz-a-pyrene is one example). No direct relationship or correlation has been established between the PCA content of an oil and its potential carcinogenicity in animals or man. Despite many studies, it can only be said that oils with very high PCA contents may well be more carcinogenic than those with low content. This lack of correlation is made more uncertain by the accelerating or inhibiting effects of other oil components, which cannot be predicted. Animal skin painting studies, with all their limitations in relation to man, are still the only practical method of assessing potential carcinogenicity, and the necessary health and hygiene precautions to be taken.

The above research programme was therefore set up to determine the extent of increase in PCA contents of a range of metal working oils during use. Whether or not such increases result in any greater hazard is, amongst other things, a question of the likely extent of exposure, and this factor must be taken into account (Vide the IP Code of Practice for Metal Working Fluids (1978)).

Thirty-three samples, provided by users and suppliers were analysed to give a representative picture of changes which may occur in use. Nine used cutting oils, five used quenching oils and five oil mist samples were selected for comparison with the corresponding unused oils. The samples were analysed by Professor Dr. G. Grimmer of Hamburg and the study was written up in the Petroleum Review.*

In summary, it was found that cutting oils used in various applications and length of time showed a PCA content increase to a modest extent. In most cases, this increase was less than ten-fold.

In quenching oil usage, PCA content increased to a much greater extent, possibly to a hundred-fold or more. This is probably due to the higher temperature involved. In a qualitative sense, it seems reasonable to assume that there could be some increased risk by

* Petroleum Review, 1981, 35, 32-33.

prolonged contact with metal working oils, although very limited contact with quenching oils would be expected. The results confirm the importance of good plant and personal hygiene practices to minimise skin contact.

Calculations from analytical data of possible atmospheric benz-a-pyrene levels which could be present in oil mist concentrations at the present TLV of 5 mgs/m^3, indicate exposures will be well below those which might present significant risk. For cutting oil sump and mist samples, the calculated benz-a-pyrene levels are below the lower end of annual average ranges for outside ambient air levels, which have been measured in various cities throughout the world. For quenching oils containing the highest level of benz-a-pyrene, the level is of the same order as the high end of these natural ranges. It is obviously prudent to maintain plant oil mist concentrations as far below the 5 mgs/m^3 level as is reasonably practical.

ANALYTICAL WORK ON N-NITROSAMINES

For some years it has been known that nitrites and secondary
amines can react to produce nitrosamines, some of which have been
found to be potential carcinogens when administered to animals in
comparatively large doses.

The apparently ubiquitous presence of these compounds either
occurring naturally or as the result of interactions, has led to the
examination of their distribution in a variety of widely used
materials including foodstuffs, beverages, cosmetics, tobacco
usage, drugs, fertilisers etc.

Water-based metal working fluids containing nitrite and commerical
tri-ethanolamine have been marketed by the oil industry for over
25 years as coolants in production engineering processes involving
metal cutting and grinding. These metal working fluids are supplied
as concentrates to be diluted with 20 to 100 times their volume of
water before use.

A review by the Committee of nitrosamines, published in the
Petroleum Review, August 1980,* confirmed that the nitrosamine
n-nitrosodiethanolamine (NDELA) could form during storage of cutting
oil concentrates containing a nitrite and an amine. Concentrations
of up to 4,000 ppm NDELA have been reported so that on dilution with
water to the normal 1-2% working concentration, the solution might
contain 20-100 ppm nitrosamine.

The employment of cutting oils at work can result in some exposure
to mist formation, although analytical methods currently available
are still to insensitive to measure NDELA concentrations in any
such mist directly.

Data have not demonstrated any evidence of a health hazard to man
arising from its presence in cutting oils. However, prudence would
dictate that efforts are made to reduce exposure to NDELA to as low
a level as possible. Unless and until satisfactory replacements for
nitrites and ethanolamines can be found, the I.P. recommend that
users should minimise exposures to fluids containing them and adopt
the appropriate protective measures and good hygiene practices such
as are given in the I.P. Code of Practice for Metal Working Fluids
(1978).

* Petroleum Review, 1980, 34, 59-60.

ANALYTICAL AND BIOLOGICAL TESTING OF USED ENGINE OIL

The study into the biological activity of used automotive engine
oils is now completed and the final report is awaited. The work
was conducted in the Shell Toxicological Laboratory at Tunstall
in Kent. This is the one large current I.P. health research project
which, on behalf of the Institute, has been carried out within the
U.K. Oil Industry. All the testhouses in the country likely to be
able to carry out such a programme were initially approached and
asked to tender, and the above Laboratory was chosen solely on its
proven capability to manage an exercise of this volume.

The objective of the study was to assess the cutaneous and systemic
carcinogenic potential of used automotive engine oils when applied
to mouse skin for a period up to eighteen months.

The oil samples tested were two unused petrol engine crankcase oils
and one unused diesel engine oil typical of market formulation. In
addition, there were three used petrol engine oils and two used
diesel engine oil specimens of the same formulations as the unused
oils, but taken after use in different types of vehicles for
different mileages.

The findings were that:-

1) treatment with the oils did not affect the survival of the
    mice, cause irritation of the skin or increase the overall
    incidence of primary systemic tumours.

2) the three unused oils (two petrol engine and one diesel engine)
    did not induce skin tumours.

3) one used diesel engine oil did not induce skin tumours.

4) two of the three used petrol engine oils were carcinogenic to
    mouse skin.

5) the other used petrol engine oil and the other used diesel
    engine oil induced a small number of skin tumours. The
    biological and statistical significance of these findings are
    still awaited.

A preliminary report of this study was published in the Petroleum
Review of October 1981.* A full report will be published in due
course.

* Petroleum Review, 1981, 35, 42.

JOINT STUDY WITH THE BRITISH RUBBER MANUFACTURERS'
ASSOCIATION (BRMA)

A joint study with the British Rubber Manufacturers' Association
into the use of aromatic extract oils in rubber compounding has now
been completed.

It was found by air sampling that ambient levels of 3-7 ring
aromatic hydrocarbons in the workplace atmosphere at the two rubber
processing units surveyed, were no higher than in the outside air.
Up to fourteen different PCA's were identified and a measured
analysis of sixteen samples was carried out in two different
laboratories using similar techniques.

Exposure to vapours from aromatic oils during use is likely to be
minimal, except at very high temperatures, but it was recommended
that they should only be used where there is no effective substitute.
The advice offered was that skin contact, especially prolonged and
repeated contact should be avoided and the other precepts - good
personal hygiene, good plant hygiene and no misuse of product should
be followed.

An Information Note on this work is to be published in the August
1982 Petroleum Review.*

* Petroleum Review, 1982, 36, 49-50.

BENZENE

Following the International Paris Workshop on the Toxicology of Benzene (1976) a number of Advisory Committee on Health members were involved in a joint Chemical Industries Association/Institute of Petroleum Committee to study further and collate the control measures for benzene in the petroleum, chemical and allied industries. The Institute of Petroleum's interest was not so much in benzene per se but in motor gasoline which may contain up to 5% benzene.

Two subsequent reports produced by the joint committee were:-

1) The "Health Precaution Guidelines for benzene exposure in the Petroleum Refinery and Chemical Industries including bulk distribution"(1980)   and

2) The "Report on Toxicology and Teratology of Benzene", which concluded that the data at present available provided no reliable basis for suspecting a teratogenic potential in benzene, the indications being that it did not.

The work of this Committee facilitated objective re-examination of the available information on benzene and together with the Evaluation by DGMK (to be reported in Dr. Kluge's following paper) was one of the factors whereby the subsequent International Vienna Benzene Workshop (1980) of experts meeting under the auspices of the Permanent Commission and International Association of Occupational Health were able to endorse the continuance of a 10 ppm benzene TLV as providing a safe hygiene standard for the industry.

CONCLUSION

With the completion of the Epidemiological Survey and the Institute
of Petroleum's three year Health Research Programme as reported in
this paper, the time is opportune for a review of future health
research activities, particularly with reference to the activities
of bodies with similar interests in the United Kingdom, DGMK and
CONCAWE (within Europe) and the American Petroleum Institute and
the Australian Institute of Petroleum.

Research projects take time, expertise and money, all of which are
not unlimited at a time when pressure continues to build up for
more regulations on health related matters. The need for a
co-ordinated approach to current issues and the best use of
scarce resources to avoid duplication, is more evident than ever
before, which is why I greatly welcome the presence of
Drs. Weaver and Joiner of the API, and Dr. Kluge of DGMK, to tell us
of the programmes carried out in their countries.

Note on nomenclature: The terms "PAH − polycyclic aromatic hydrocarbons" and "PCA − polycyclic aromatics" used later in these papers
are synonymous. In an earlier period they have been referred to as
"polynuclear aromatics (PNA's)".

AMERICAN PETROLEUM INSTITUTE RESEARCH
ACTIVITIES, AND U.S. ASPECTS OF REGULATORY
SCENE CONCERNING OIL EMPLOYEE AND PUBLIC
HEALTH

Dr. R.E. Joyner

Chairman, Medicine & Biological Science
Department, American Petroleum Institute,
and Corporate Medical Director, Shell Oil
Company, Houston

1982 marks the 24th anniversary of cooperative biomedical research
conducted for members of the U.S. oil industry by the American
Petroleum Institute. That research had its beginnings in 1958
when the use of petroleum waxes in food packaging was threatened
by one of the premier occurrences of the now ubiquitous "cancer
scares". In response to that threat, a group of API wax producers
funded a study by Dr. Shubik, then at the Chicago Medical School,
which demonstrated that feeding, skin application, and even
subcutaneous injection of food grade waxes gave rise to no adverse
effects. As a result, the waxes continued in use for a number of
years until economics and developing technology produced other
substitute materials which displaced those waxes from the
marketplace.

Twenty years ago studies on certain high boiling aromatic fractions
derived from catalytic cracking were sponsored by API and conducted
at Kettering Laboratories. These materials were shown to be
carcinogenic in animal tests, and these results stimulated the
subsequent development of appropriate precautions and safeguards
designed to protect employees and customers from potential exposures.

In succeeding years, the API sponsored a series of literature
reviews by experts in various fields which focused on the toxic
properties, safe handling procedures, and treatment of adverse
reactions to some 17 materials used in the industry, among them
aromatic petroleum naphtha, aromatic oil, benzene, kerosine,
naphthalene, sulfuric acids, toluene and xylene.

By 1968 the menace of arbitrary and potentially capricious
regulations of petroleum products was growing rapidly. It became
clearly apparent that only modern, soundly based toxicological and
exposure data would be of any value in forestalling unwarranted
regulation. A 5-year program involving a matrix study of typical
hydrocarbon solvents was begun at Carnegie-Mellon Institute which
ultimately cost $600,000. Although the original goal of developing

a formula to predict toxicity on the basis of boiling range,
C–number, or aromaticity/paraffinicity ratio was not attained, some
16 scientific publications resulted which provided the Threshold
Limit Value Committee on the American Conference of Governmental
Hygienists (ACGIH) sorely–needed evidence upon which to establish
rational occupational exposure limits to these materials. These
carefully designed and executed experiments clearly demonstrated
that there was no cheap or swift methodology in these areas that
would allow one to derive valid scientific data. The slow, costly
process of animal bioassay was established as the method of choice,
and the API was designated as the organization which could perform
these studies for the member companies in a cooperative and cost–
effective manner.

Subsequent studies have been expanded to benzene, motor gasoline,
crude oil and its major fractions, shale, shale oils and spent shale.
In addition to basic studies of acute toxicity resulting from dermal
application, feeding and inhalation, chronic lifetime studies of
carcinogenic potential of a number of compounds have been carried
out via the skin, gastrointestinal and respiratory tracts. More
recently, clearly defined lines of investigation have been developed
in response to the growing interest in such areas as mutagenicity,
reproductive effects and neurotoxicity.

The rapid growth of research activities within the past 5 years has
prompted a thorough reappraisal of our objectives and the
development of specific guidelines with respect to just what kind of
research API should do – and should not do.

It is currently accepted that the primary role of the Medicine and
Biological Science Department of API is to plan, support, monitor
and evaluate health–related research in an effort to ensure that
petroleum operations do not adversely impact employees, customers,
the public, or the environment. In fulfilling that role, we are
also cognizant of the need to spend some of our expenditure in
methods development and basic research for new techniques, in order
to be able to supply the most cost–effective and time–effective
methodology to our overall program.

In a quantitative sense, the growth of API biomedical research can
best be appreciated by a brief examination of the budget allocated
in this line of endeavour. In 1972 $350,000 was budgeted for
various studies conducted by the Institute. Four years later, in
1976, this figure has risen to $1,350,000. Most of this increase,
incidentally, represents real growth, not inflation. The budget
for the current year is $4,900,000. In sum, the Institute has
budgeted in the past 10 years more than $21,000,000.

It is certainly not unexpected that expenditures of this magnitude
should draw considerable attention from the officers of member
companies, especially in today's economic climate and in view of some
fiscal retrenchment which is evident internally within most of the
companies. These factors have resulted in a budget reduction of
approximately 20% for the calendar year 1983 program.

Currently there are underway 40 specific projects involving the
disciplines of Occupational Medicine, Toxicology, Industrial
Hygiene, Environmental Biology and Analytical Chemistry. Time
will not permit more than a cursory look at a program of this
magnitude. I would like to mention, however, 3 of these projects
which were selected on the basis of their potential significance
to the industry. These are:-

ITEM 1:   A study of the neurotoxic potential of n-hexane in the
          presence of other hexane isomers. (PS-29)

The determination of the neurotoxicity of selected petroleum samples
is an important objective of the API research program. Adverse
effects upon the nervous system are currently categorized by the
Environmental Protection Agency as among the more serious health
effects, knowledge of which must be reported to the agency under
the provision of the Toxic Substances Control Act by companies
who became aware of such effects.

Another measure of the importance of this line of research lies
in the fact that for certain chemicals the manifestation of
neurotoxicity is the primary determinant used in setting permissable
occupational exposure limites. Exposure limits set too low result
in needless capital expenditures on engineering controls. Set too
high, of course, they can lead to serious, and possibly irreversible,
health effects. Clearly, we need definitive and reliable data upon
which to base such important regulatory decisions.

This study, which has recently been completed, was designed to
determine whether or not the established neurotoxicity of n-hexane
is augmented by concurrent exposure to the hexane isomers commonly
encountered in commercial grade hexane solvents. Dr. Peter Spencer
of the Einstein School of Medicine is the principal investigator.

Rats were exposured to varying levels of n-hexane and n-hexane with
varying levels of isomers for a period of 6 months, 22 hours/day,
7 days/week. While the n-hexane produced the expected neuropathy
at levels of 500 ppm, there was no evidence that the concurrent
presence of isomers, even up to levels of 500 ppm, caused any
potentiation of this effect.

The importance of this finding to our industry is that is supplies us
with unmistakable evidence that the presence of isomers with
n-hexane poses no greater health risk than that of n-hexane alone.
This evidence can be used to assist in refuting one currently-held
postulate that the present permissable exposure level for n-hexane
should be lowered in order to protect against a presumed
enhancement of effect due to concurrent exposure to isomers.

ITEM 2:   A study of reproductive hazards associated with exposure
to benzene (PS-39)

"Reproductive Risk" is a catch-phrase that is becoming very popular
in the U.S. Some scientists believe that potential reproductive
hazards constitute as great a challenge to researchers as does
cancer. Sterility, spontaneous abortion and birth defects have all
been attributed, among other causes, to exposure to certain chemical
substances. There is a need to determine the potential effects of
selected high-volume petroleum products and intermediates on all
aspects of reproductive performance. Well-designed research
addressing both female and male reproductive effects has become so
costly that a collective effort is essential.

This study, also recently completed, was designed to evaluate the
potential teratogenic and/or fetotoxic effects of exposure to
benzene. The study was conducted at Hazelton Laboratories of
America, Inc.

Four groups of pregnant Sprague-Dawley rats (40/sex/group) were
exposed to benzene vapor at concentrations of 1, 10, 40 and 100 ppm.
Exposures continued for six hours daily from days 6 to 15 of
gestation. Two groups of pregnant rats (40/sex/group) were exposed
to ambient air only and served as control groups. Criteria evaluated
in the animals included maternal mortality and morbidity, clinical
observations, body weights, gross pathology, implantation efficiencies,
and fetal viability, size and development.

There were no statistically significant differences between the
two control groups on any of the parameters mentioned.

Mean fetal body weights in both sexes of the group exposed to 100 ppm
of benzene were significantly decreased when compared to each
control group. Thus, a slight fetotoxic effect was observed at
100 ppm. This effect was not observed at the 40 ppm level. No
treatment-related effects were noted in the remaining maternal and
fetal data generated by this study. Based upon these findings,
the test material, benzene vapor, was considered not to be a
teratogenic agent to Sprague-Dawley rats at 100 ppm or below.

In addition to benzene, numerous other industry substances of a
generic nature have been tested in previous API studies to
determine their potential for adverse reproductive effects. These
studies, all of which have failed to demonstrate any reproductive
hazard, have included unleaded gasoline, xylene, toluene, stoddard
solvent, hexane, and kerosine, among others.

The importance of these findings to industry is obvious, given the
rising societal concerns regarding reproductive effects, and the
increasing attention being paid to this subject by litigating
attorneys in the U.S.

ITEM 3: A study to evaluate the chronic inhalation toxicity of
         unleaded gasoline in mice and rats

Cancer is the second leading cause of death in the U.S. today.
Because of this, it is also the health issue of greatest concern to
our industry. A significant percentage of cancer has been attributed
by various authorities to environmental causes. Although exposure
to industrial chemicals may contribute in only a minor way to
environmental cancer causation, their role as causative agents
must be more fully defined since appropriate control of these risks
could make a positive contribution to cancer control. The potential
carcinogenicity of petroleum products, base stocks and intermediates
must, therefore, be established as clearly as the current state-of-
the-art will permit. Toxicological and epidemiological investigations
are essential in determining human cancer risks, particularly those
due to chronic, low level exposures to chemical substances.

In addition, the possibility of a provable relationship between
mutagenicity and cytogenetic abberations on the one hand and
subsequent development of cancer on the other must be
painstakingly evaluated with reliable studies.

This current study of unleaded gasoline has been completed, and a
final report has been submitted to the government agencies. The
contractor for the study was International Research and
Development Corporation.

This investigation was designed to determine the chronic toxicity,
in rats and mice, of long-term exposure to wholly-vaporized unleaded
gasoline containing 2% benzene, as well as a hindered phenol
antioxidant and an amine metal deactivator. 800 Fischer 344 rats
and 800 B6C3F mice with equal numbers of each sex were exposed to
either 67 ppm, 292 ppm or 2,056 ppm of the wholly-vaporized gasoline.
Additional groups, exposed only to ambient air, served as controls.
Exposures were for 6 hours/day, 5 days/week for two years or more.
All interim and terminal sacrifices were subjected to complete
histopathological examinations in accordance with guidelines of the
National Cancer Institute.

The exposure phase of this study was completed in November 1980.
On November 14, 1981 the contractor reported that histopathological
study of tissues from the terminal sacrifice had revealed that renal
carcinomas were present in some of the aged male Fischer rats
exposed at all three dose-levels. These observations have been
subsequently confirmed by a consulting pathologist from
Experimental Pathology Laboratories, Inc.

Specifically, the findings were as follows:

(1) In the control group (non-exposed) no proliferative renal lesions
     were observed.

(2) In the group exposed to 67 ppm, two renal carcinomas were
     recorded.

(3) In the group exposed to 292 ppm, two renal carcinomas were agreed upon by at least two pathologists. There was disagreement between the two pathologists as to the existence of several additional undifferentiated or mixed malignant tumors.

(4) In the group exposed to 2,056 ppm, five renal carcinomas were observed and corroborated by the consulting pathologist.

It should be emphasized that all of the renal tumors described occurred only in male rats with the exception of one renal sarcoma observed in a female rat exposed to 292 ppm. None of the exposed mice developed tumors of the kidney.

It is also noteworthy that the renal tumors were observed only in aged rats – those surviving the 2-year period of exposure – a fact which has obvious bearing on any assessment of potential risk to humans. This observation, along with the fact that only one species and essentially one sex of that species developed a tumor response clouds considerably the extrapolation process with respect to human risk. On the other side of the coin, the demonstrated dose-response observed in the male rat cannot help but raise one's level of concern.

Of obvious importance in this study is the fact that a no-effect level for the observed renal carcinomas was not established under the protocol used. The question which faces us, now, is simple: Where do we go from here?

I remind you that in this study animals were exposed to wholly-vaporized gasoline, an exposure quite unrealistic in terms of "real-life" human contact with gasoline vapor. This fact must be given great weight in decisions regarding any further animal studies; and such studies, if conducted, would utilize vapor exposures which more closely mimic the real-life situation.

A special task force has been appointed within API to consider what additional studies may be indicated in light of these findings. One approach under consideration is a thorough review of existing epidemiological studies on human cohorts possibly exposed to gasoline vapors. Additionally, some efforts may be directed toward mounting a new epidemiology study to assist in evaluation of these animal findings. A proposal has already been made for funding of a feasibility study in this regard.

Winston Churchill is quoted as having said, "True genius resides in the capacity for evaluation of uncertain, hazardous, and conflicting information." This particular piece of research, without doubt, presents splendid opportunities for persons with such talents! The only firm conclusion which can be drawn at the moment is that no firm conclusions can be drawn with respect to human risk.

I would like to turn now to some comments on recent regulatory
activities in the U.S. which significantly impact our industry.

Actually, in the past 5 years concern over regulation in the U.S.
has been more or less muted by an even greater concern over a much
more ominous threat - that of litigation. There are now literally
thousands of lawsuits alleging personal injury from industrial
products. In the asbestos industry alone, as of one year ago, more
than 12,000 suits had been filed and new ones were being recorded
at a rate of 400/mo. The Johns-Manville Co., the giant of the
asbestos industry in the U.S., has been reported to be the defendant
in more than 8,000 of these suits. In sum, these actions represent
the largest and potentially most costly block of product liability
claims ever to confront American industry.

Positive research test results have been, and will continue to be
used as "admissions" by industry of the hazard associated with its
products - even though particular study results were not designed
for, or intended to demonstrate, the degree of human risk associated
with the substance or product under test.

Notwithstanding the risk of misrepresentation and misuse of its
research results, industry has no choice legally but to continue
health and safety research. The law clearly places on industry the
duty of an expert to know and warn of hazards associated with its
products. The continuing demonstration of good faith and a track
record of honest effort will be industry's greatest asset in
defending it's research against judicial misunderstanding and abuse.

Two years ago, July 2, 1980, the U.S. Supreme Court issued its
decision in the Benzene Case. As you will recall, the Occupational
Safety and Health Administration (OSHA) had proposed lowering
the permissible exposure level from 10 parts per million (ppm) to
1 ppm and had been challenged by industry in the courts. The
Supreme Court ruled, in part, that OSHA had failed to meet it's
burden of finding a significant risk of material impairment of
health to justify it's action in lowering the permissible
exposure.

OSHA learned an important lesson in the Benzene Case, one which is
giving the Administration pause for thought before future health
standards are proposed. The decision, however, did not dissuade
the then Assistant Secretary, Eula Bingham, from a final blast of her
shotgun. In the two months between the election and inauguration
of Ronald Reagan, OSHA initiated nine rule-making edicts which have
since become known as the "Midnight Regulations." These, along with
6 other proposals, were left on the table to be dealt with the
incoming Republican administration.

The new OSHA, under the direction of Thorne Auchter, has been in
place for 18 months, now, and has addressed the Bingham legacy as
follows: one standard was put into effect (electrical safety); two
were partially stayed (lead, hearing conservation); four were
withdrawn (labeling, walk-around pay, generic carcinogens amendment,
withdrawal of Indiana's state plan); and eight are in abeyance

(conveyors, hazardous materials, marine terminals, and five others initated earlier in the Bingham administration, including the generic pesticide standard, generic reproductive hazards, chromium, confined space entry, and lockout/tagout).

The Auchter OSHA, to date, therefore, has been faced with the time-consuming chores of reconsidering the unrealistic legacy of the past, as well as developing proposals of their own in the area of protection of worker health. Among the new initiatives worthy of mention is the subject of Health Inspections. Industry is now urging, (and many believe OSHA is considering) a new approach to health inspections. The previous practice of targeting health inspections to facilities which are identified as handling hazardous substances has in many instances proven extremely wasteful of manpower and money. This has resulted because of OSHA's failure to give any pre-inspection consideration to the facility's manner of handling hazardous substances, ie., the degree of sophistication of that facility's programs in Occupational Medicine and Industrial Hygiene. The new approach being advocated would involve pre-inspection assessment of these programs in an administrative review. If the facility met certain criteria, as yet unestablished, the inspection might be concluded in 1-2 days rather than the usual several weeks. Exploration of this concept is proceeding, as are other alternative proposals with respect to achieving a safe work-place through voluntary programs initiated by the employer.

Perhaps the largest regulatory problem looming on our horizon is embodied in the current administrations' philosophy of reducing federal regulations of all kinds and turning regulatory authority back to State Governments. In many areas, there is much to commend this; there are, however, areas where a Federal regulation, if a feasible one, can prevent chaos. This is especially true of areas where the crossing of state lines, as in product safety labelling, is involved. A set of regulations killed in Washington can conceivably reappear as 50 sets of regulations, with differing provisions, in the 50 States. For large companies which operate in multiple states, the trend toward creation of multiple and differing regulatory administrations has already become noticeable and even onerous, most particularly in the State of California.

To cite one recent example of these difficulties: The work of two British Scientists, Dr. Robert Murray and Dr. Kevin Browne, in reviewing two current animal studies involving asbestos has been published in a recent issue of Lancet (August 22, 1981, Vol.2 (8243) p415). One of these studies, conducted by Bolton, Davis and Lamb of the Institute of Occupational Medicine in Edinburgh*, "confirmed the absence of asbestos-related gut tumors in a large-scale study in rats monitored for their whole life span." The other study, conducted by the National Institute of Environmental Health Sciences in the U.S. has been reported by the U.S. EPA as showing that "the initial results did not detect any adverse effects on the test animals from asbestos in the diet." Drs. Browne and Murray concluded that the evidence from animal studies and epidemiology "is beginning to point strongly away from any

association between asbestos fibers and an excess mortality from
gastrointestinal tumors."

Despite this evidence, the State of California has recently mandated
that exams for asbestos workers should include a clinical
examination of the abdomen and a laboratory test for occult blood in
the stool. Needless to say, such regulations will add substantially
to the cost of medical surveillance without reasonable expectation
of significant benefit to the workers. Given the current economic
climate, none of us can afford regulatory actions which increase
costs and impinge upon productivity in the absence of offsetting
benefits to the health or safety of our employees.

Auchter has stated publicly that his OSHA administration should be
judged by whether or not an improvement in workplace safety and
health occurs during this four-year period. At issue is whether
Auchter's approach will achieve more than did the police-type
enforcement of the previous administration. It is an important
question, and a matter that rests squarely on the shoulders of
employers.

There is a very real concern on the part of many that employers may
be apathetic to the OSHA reform now underway, perceiving it as a
fortuitous interlude instead of as an opportunity to demonstrate
that real improvement in safety and health is best achieved by the
approaches of the current administration, and not by the excesses
of the past.

* 'The pathological effects of prolonged asbestos ingestion in rats',
  in press Env. Research.

APPRAISAL AND COMMENT ON DAY 1

Sir Richard Doll

Regius Professor of Medicine, Warden of
Green College, Radcliffe Observatory, and
Hon. Director, ICRF Cancer Epidemiology
and Clinical Trials Unit

It is a pleasure to make some comments on the very interesting
material we have had presented to us but a requirement to make an
appraisal of everything that has been said is really a very major
one, which I cannot rise to.

The conference today is, I think, particularly well timed for two
reasons. One has been referred to frequently already, namely the
increasing realisation of the need for detailed and reliable
evidence of the effect of environmental hazards associated with any
major industry as they actually affect man. This is usually seen
most easily in the employees of the industry, but sometimes, as in
the case of tobacco, it is more easily seen in the people that use
the products. The second reason has not been referred to often,
but it does make some of the observations that have been reported
today of particular pertinence, and that is the potential hazard of
the switch from petrol to diesel engines throughout the world. The
question of the possible effect of diesel exhausts is not a new one.
I am sure you have discussed it and heard it discussed on many
occasions in the past. I heard it first raised about thirty-five
years ago, and I am quite certain that we are going to hear a lot
more about it in the future. I will, therefore, refer to it in
my concluding remarks.

The papers that have been given provide a very valuable survey of
many aspects of the environmental effects of petroleum and of the
use of petroleum and petroleum products. Some of the papers are
so complete in themselves, being given by a central figure in the
field with which they are concerned, like Dr. Duncan, and others are
so encyclopedic, like the surveys reported by Dr. Kluge, that it is
impossible to add anything to them. Other speakers have touched
briefly on such complex problems as Dr. Leese's reference to the
effects of lead in air from lead in petrol, that it would be quite
improper to try to say anything about them in the course of half an
hour. I shall, therefore, make very few remarks about these
papers. I am, however, sure that I speak for everyone here when
I say that I greatly welcome the very detailed measurements that
the German industry is making of the extent of exposure to benzene in

89

service stations and in other parts of the petroleum industry,
as they will be of great interest to many people. I also welcome
the recognition of a place for epidemiology in Germany. Many parts
of Western Europe, to our regret have been a little slow in
recognising that epidemiology has any place to play at all in
determining the existence of environmental hazards or in measuring
their effect when it was demonstrated they did exist. I greatly
welcome this recognition that epidemiology has a part to play,
not only in measuring a harmful effect or checking that the effect
has been eliminated, but also in making sure that new substances
that are introduced into industry are safe to handle even when no
hazard has been demonstrated by toxicological investigation. One
only has to think of some of the drugs that have been introduced
in the last ten years and have had serious effects when used, that
appeared to be quite safe toxicologically and even after the human
effects were demonstrated, have not been shown to be hazardous in
animal experiments; for example, Practolol in England and Aminorex
in Germany.

For the rest, I shall concentrate on Dr. Alderson's papers, partly
because he was kind enough to provide me with the detailed reports
beforehand and partly because some of the general issues raised in
them are of relevance to all the epidemiological studies that
are being carried out by the industry throughout the world.

Dr. Alderson has reported the results of these very substantial and
important studies. I have only a few trivial criticisms of the
techniques involved and it is not surprising that I should have
some, as no two specialists would ever discuss investigations of the
size of his without finding some points to disagree on, but as they
are very technical, I shall omit them. His papers do, however, raise
one question of interpretation that I think we should spend a little
while discussing, namely the acceptability of the so-called healthy
worker effect as explaining a low standardized mortality rate and
the impact of this on the interpretation of a normal standardized
mortality rate for malignant disease.   This is an
important practical problem and I was delighted to hear from
Dr. Joyner that McMahon and Monson, two epidemiologists whom I
greatly respect, are going to examine this problem in depth and write
a report on it.

Now the three studies that Dr. Alderson reported had overall
standardized mortality ratios of 84% in the 8 UK refineries, 85%
in oil distribution workers, and 85% in the London Transport
Maintenance group, based respectively on 25 years follow-up, 25
years follow-up, and 8 years follow-up. To me, standardized mortality
ratios is the middle 80s are quite acceptable as rates that can
be attributed to the healthy worker effect. But we do need to
demonstrate that a deficiency of this order is in fact due to a
healthy worker effect. In fact, if you are working for an
insurance company, it is extremely difficult to predict what the
risk of death of an individual is going to be in 20 years
time. If this were not so, insurance companies would be doing even

better than they are now. The healthy worker effect ought, therefore, to work off in the course of time. Some years ago it was said that it should work off in about five years. More recently, Fox, in this country, has said that it takes nearer fifteen years before it works itself off and it may never disappear completely. We do, however, like to see, in a report that explains a low mortality as being due to a healthy worker effect, SMRs for different periods after each individual came under observation: for example, for the first five, five to ten, ten to fourteen, fifteen to nineteen years etc., after the individuals came under observation. One is much happier at attributing a low mortality to a healthy worker effect if the SMR is found to be gradually approaching 100% over these periods. Alternatively one has to consider the possibility that the SMRs are artefactually low and there are at least two ways in which they could be artefactually low in the refinery and distribution studies. In the reports on both these studies, comments were made to the effect that the deficiency of deaths was greatest among the men who were employed earliest. That is a slightly odd finding, because if there were any hazards one would expect them to be greatest in the men who were first employed, longest ago, when the industrial conditions are likely to have been worse than they are now. One wants to know, therefore, why this occurred.

One way in which this could have occurred artefactually was mentioned by Dr. Alderson, and that was the failure to determine all those who had emigrated. This will have particularly affected the follow-up study in which the Social Security records were used for tracing because all they tell you is whether a person has died or "we have no record of his death". They will not tell you, for reasons of confidentiality, that the man or his employer has actually paid a contribution for him in the last month and that he can, therefore, reasonably be thought to be alive. All they will tell you is that they know he is dead, as they have paid his funeral benefit or that they do not. They do not know if he has emigrated, and people who were first employed in the 40s are much more likely to have emigrated than the people who were first employed in the 60s. By keeping the people who have emigrated (but are not known to have emigrated) in the population one obtains an artificially high expected mortality or alternatively an artificially low observed mortality whichever way you like to look at it. People who were first employed in the 40s, have to be still in employment in January 1950 when the studies started, or they would not have been included; but the possibility of emigration in the early 50s can not be ignored.

Then there is another possibility, which is a rather more difficult to re-assure oneself about, which always arises when records are collected retrospectively, and all these studies are retrospective cohort studies in character in that the population at risk was identified as having been in employment some years in the past. In some such situations there is a possibility of preferentially omitting the records of people who have died. By saying this, I am not wishing to suggest that they are omitted intentionally (although in certain circumstances this could happen).

But it can easily happen that the records of people who have died
are destroyed by somebody who says "That record is of no interest,
the man has died and nobody will ever want to see that again".
Alternatively the record can be of such interest that it has been
examined by many people and eventually lost. I know of several
examples in which retrospective cohort studies have shown an
artificially low mortality for people who entered the study a long
time ago, because of the differential disappearance of records of
people who had died. It depends, therefore, on how you build up
your population. Whether you can be confident that bias of this
sort has or has not crept in depends on the way in which the study
population was constructed. There is not likely to have been a
large bias from this source in these studies, but one would have
liked to see data separately for different periods for people who
were first employed in different calendar years, just as one would
have liked to see mortality rates for different periods after first
employment.

If we now consider neoplasms, it is notable that these three studies
showed standardized mortality ratios of 89, 87 and 95%, corresponding
to the overall mortality ratios of 84, 85 and 85%. I am much less
happy to accept a standardized mortality ratio under 90 for
neoplasms as due to a healthy worker effect. For I have no idea
how to predict that a man or woman is going to get cancer in ten
years time unless, perhaps, you can tell me what his smoking
habits are or how much alcohol he drinks. If the subject is a
woman I can predict the risk to some extent if you tell me how old
she was when she had her first full-time pregnancy or how many
sexual partners she has had or even how many sexual partners her
husband has had (which is relevant to the development of one type
of cancer in women). But these are not the sort of data that are
normally recorded when you first employ people, and they have
certainly not been taken into account in these studies. I am much
happier with standardized mortality ratios of the order of 100% for
all neoplasms when you have taken the trouble to do what Dr. Alderson
did, namely adjust for geographical region of employment.
(Incidentally, I would agree that he was right to adjust only for
geographical regions and not for social class for the reasons he gives
in his papers).

If you look in more detail at his data for neoplasms you find that
in the two first studies, where the SMR for all neoplasms was below
90%, there was a particularly low mortality for cancer of the lung.
This needs explaining. Is it conceivable that the employees
in these two groups smoked less, because we know of no other way
in which we can predict people who have an abnormally low
mortality from lung cancer? There are several smoking-related cancers
including, in all probability, cancers of the bladder, pancreas and
kidney as well as cancers of the lung and upper respiratory and
digestive tracts (although not such a high proportion of the former
types of cancer are due to smoking as of the latter). If now you
class all these types together and take them from the total and
look at the remaining cancers, then what you find is that the

standardized mortality rate from neoplasms other than those
related to smoking in the refinery population was 99%, in the
distribution group 95%, and in the garages 87%. The number of
deaths in the last study was a good deal smaller so the rate might
have been materially affected by random variation; moreover the
expected mortality was not adjusted for region. I am quite happy
with figures of 99% and 95% in the first two studies but to reach
them we have to conclude that the populations of employees were in
fact smoking less than the national average and I hope that someone
in the course of the discussion will be able to say whether or not,
in their opinion, these groups of people did smoke less, because
of restrictions on smoking due to the nature of their work.
Certainly in the garages, where smoking related cancers were not
less frequent than normal, restrictions on smoking were removed in
1950 when 85% of the buses already had diesel engines, so we know
that the one group of employees that did not have a low mortality
for smoking-related cancers worked in an area where there were no
restrictions on smoking - but was the converse true for the others?

It follows from what I have said, that I do not think that one
should regard a standardized mortality ratio from neoplasms
of 100% as indicating a hazard just because the death rate from
other diseases is about 80 or 85% of normal. This combination is
what you must expect to find in an epidemiological study of any
large industry that requires any degree of physical competence in its
staff.

If we now look in more detail at specific conditions, it does not
appear that the refinery data point to any hazard. If we test
the a priori hypotheses that Dr. Alderson listed, I do not think they
gave any support for the idea that there should be any excess
mortality from gastrointestinal cancer. The data on nasal sinus
cancers are interesting; there were only seven deaths from this
disease as against 3.0 expected but four of them occurred at the
location which Dr. Alderson described as "J". Some years ago, in an
independent study of the ratio of male to female deaths from nasal
sinus cancer in Britain I noticed that the ratio was unusually
high in this area and I made some enquiries from the local Medical
Officer of Health to find out what the occupations were of the men
who had died of nasal sinus cancer. We failed, however, to obtain
any useful information. In Dr. Alderson's study only two of the
four men who died of this condition were long-term employees, two
had quite trivial periods of employment and I think there was a
statement that the company had taken over a chemical company that
had previously been operating there. There may well have been a
hazard of nasal sinus cancer in this particular town, but it need
not have anything to do with the petroleum industry. I hope that
can be looked into a bit more closely.

Blot's suggestion, based on his study of the geographical distribution
of cancers in the U.S., that there might be an excess of skin cancer
associated with the industry did, I think include both melanoma
and squamous carcinoma of the skin. Melanoma is one of those few

cancers that are becoming more common throughout the world. In all white-skinned populations not only the incidence but also the mortality from the disease has increased substantially over the last thirty years. It has not increased in black-skinned populations. Now this is probably attributable to the greater exposure of the skin to ultra violet light, something which is readily acceptable to members of my generation who have seen the changing size of bathing costumes. However, scientifically, the relationship is not all that clear, partly because of the sites on which the melanomas appear and partly because the greatest increases have occurred in Scandinavia, which may, perhaps, be explained by the fact that so many Scandinavians go to Spain for their holidays. The consensus of opinion is that the increased incidence is due to greater exposure to ultra violet light, but a question must lurk at the back of one's mind about the possibility that an industrial hazard has contributed to it, and it was therefore, important to look for an industrial hazard by carrying out an intensive study of the occupational history of the affected men.

As for lung cancer, which has also previously been postulated as a hazard in the industry, the studies very clearly show that there is no hazard; nor does there appear to me to be any support for the idea that exposure to petroleum produces cancer of the brain, lymphoma, or leukaemia.

The distribution workers' study for its part, generates an a posteriori hypothesis that there may have been a special risk of ischaemic heart disease, the mortality ratio for this disease being 99% as opposed to ratios of around 80% for most other non-neoplastic diseases. Clearly, we must ask why this occurred. The obvious explanation, that it is something to do with the drivers, who have been shown in other studies to have increased mortality from ischaemic heart disease, is not borne out, because the relatively high ratio is approximately equal in all the occupational groups. There is no evidence that it is occupational in origin and I suspect it is due to some aspect of life-style. Could it I wonder be due to a relatively small use of alcohol? One of the surprising things that epidemiologists have convinced themselves of in recent years is that the lowest mortality rates in relation to alcohol consumption do not occur with the life-long abstainers, and even total abstainers amongst epidemiologists agree that this is in fact so. The lowest mortality rates occur in a group of people that drink the equivalent of about two or three glasses of wine a day and it could be that the employees in this group for some reason drink relatively little alcohol. It is notable that their mortality from cirrhosis of the liver was slightly under 50% of that experienced by the general public which provides some support for the idea that they do not drink as much as average. However, this is a very speculative explanation which should not be taken too seriously without some stronger supporting evidence.

As for leukaemia, there is not, to my mind any evidence in this study of an excess mortality in any subgroup. One possibly important bit of evidence that has not been referred to so far concerns the possibility that some of the cases may have been classified as cases of erythroleukaemia. In cases attributed to benzene exposure a substantially higher proportion of leukaemias are recorded as erythroleukaemia than in the general run of cases. Normally the proportion is about 1%, but it is something like 10-20% in cases attributed to benzene which otherwise are nearly all myeloid leukaemias. Erythroleukaemia is not referred to in the report. It would presumably have been categorized with the unclassified leukaemias and it is notable that there was no excess of unclassified leukaemias. If there were even one case of erythroleukaemia in the employees, I would be more inclined to think that there may have been a benzene hazard. As the evidence stands, I think the possibility of a hazard can be dismissed.

The interpretation of the case-control study of leukaemia to which Dr. Alderson referred is I think extremely difficult. I do not like picking out sub-groups as showing an excess when the overall population does not show an excess of a disease, unless the excess is gross and occurs in some small and accurately defined group. In this instance, the small excess in the case-control study affected about one-third of the population and was compensated for by a deficiency in the rest, and when one has that situation and an overall mortality no higher than normal, it is just as reasonable to interpret the deficiency in the other group as being abnormal as to attribute the excess in the "medium-high exposed group" to their occupation.

A hazard of lymphoma is a little more difficult to dismiss. Goldstein's review referred to a variety of types of tumour in the lymphoma series which has been reported as being in excess in the industry and there is a small excess in the distribution worker study of the broad group of lymphomas, including myeloma and myelofibrosis. A year ago, I would have dismissed that on the grounds that you do not get agents causing a whole range of lymphoreticular cancers, including both non-Hodgkin's lymphoma and Hodgkin's lymphoma, myelomatosis, and leukaemia, but I am less happy to do that now in the light of the observations that have been made on this group of diseases in workers exposed to certain herbicides. This is a very contentious and difficult issue but I have personally come to the conclusion that their exposure probably does produce a hazard and one therefore now has to consider the possibility that agents may cause this group of diseases as a whole and not just one type of them.

The London Transport garage maintenance worker study was essentially negative and this is very important to the general problem of assessing the effects of diesel fumes. The problem is epitomized in a paper published by Schenker (1980) in the United States which states that diesel engines contribute practically all the polycyclic hydrocarbons in the air that are produced by automobiles and that petrol engines produce practically none. It follows that a switch to diesels would involve increasing pollution with polycyclic hydrocarbons in the air. Schenker also said that the data from Britain, which had been referred to by Raffle's study of bus garage workers was irrelevant because the men had not been exposed long enough, and had previously been exposed mainly to petrol fumes. Confusion has, I think, arisen because observations on exhaust fumes in America relate to the post-1978 period when petrol-engined automobiles coming into the market were required to have filters to get rid of the carbon monoxide and these also incidentally got rid of polycyclic hydrocarbons. In the United Kingdom, according to Drs. Waller and Lawther, diesel motors have produced no more polycyclic hydrocarbons mile for mile than petrol driven motors and if this is so, the epidemiological experience of people exposed to motor exhausts in this country, including the bus garage employees that have been referred to, can be taken as evidence of the effect of all automobile exhausts, including diesel exhaust. Measurements made in the bus garages now show that the amounts of polycyclic hydrocarbons to which individuals are exposed from diesel exhausts are absolutely minute (Waller, 1982). Twenty years ago they may have been about 12 nanograms per cubic metre, which is only a very little more than occurs generally in the outside air and is quite a trivial amount. I would have thought that the interpretation of all the data we have been presented with, and a great deal more on the epidemiology of lung cancer in general, is that diesel fumes, even if all motor cars in the world were to become diesel driven, would not add any material risk of lung cancer to the population. I would like to leave some minutes for discussion so I will conclude just by making three more comments. One on Dr. Weaver's very detailed review of all the reported studies, including several unpublished ones. I thought he showed convincingly that epidemiologically speaking there was no reason to suppose there was any excess of cancer among workers in the petroleum industry. The hypothesis-generating study based on proportional mortality ratios to which he referred is not to me an acceptable one. That is not to say that proportional mortality ratios are never of any value. If they just pick out one disease quite sharply distinct from all others, then this can be of great importance. But if you accept, as I think we should, that the healthy worker effect reduces the mortality from most diseases other than neoplasms, but not from neoplasms (at least not after a few years), then inevitably a proportional mortality study is going to show excesses of several neoplasms and this is exactly what was found in that study. In other words the excesses were a reflection of the healthy worker effect.

The cohort studies that have been carried out fail to provide any consistent evidence of any specific excess. They will all inevitably show some excesses, but there is no consistency between them and, apart from the two point which Dr. Alderson said should be investigated, we have I think, no further reason to worry about occupation hazards of cancer in the petroleum industry in Britain.

In conclusion, I should like to express my sympathy with Dr. Duncan when he talks about the social hazards of conducting epidemiological studies. I entirely agree with him that there is a danger of their being misinterpreted. But there is also a danger of misinterpretation if epidemiological studies are not carried out. Indeed, I believe that this danger is even greater, when the question is asked "why are they not being done?". The only solution to my mind is education. I believe we must have confidence in our ability to educate people about the interpretation of this type of scientific study. In this spirit, a colleague of mine wrote to the Editor of the New Scientist a year or so ago when an editorial stated that a trade union had had a conference on occupational hazards of cancer and had been unable to get any UK scientist to participate in it and so had had to call on Epstein from the United States. He enquired what attempt had been made to get UK scientists in the UK to participate, because we were not aware of anybody having been approached. The upshot was that we subsequently, at the request of the Union, ran a symposium on the uses of epidemiology in the detection of occupational hazards specifically for the safety officers of the Union and ever since then have had regular letters from the Chief Safety Officer of the Union asking for our advice on the interpretation of some new paper. I welcome this and think that if such relationships can be developed, we should over the course of years, be able to build up an educated body of safety officers in the Unions and in this way overcome the risks in which Dr. Duncan very understandably referred.

I would also like to say, in relation to Dr. Joyner's remarks, that I recognise that epidemiology is not the weapon of choice for detecting hazards. We must, however, learn much more about the causes of disease and particularly the actual mechanism by which cancer is produced, before we can sensibly interpret very many animal and in vitro experiments and assess what they actually mean to man. At the present moment, it is easy to write a long book on what they mean in general, but to be precise about what they mean in a particular case is often impossible.

REFERENCES

SCHENKER, M.B. (1980)   Diesel Exhaust – an occupational carcinogen? J. Occup. Med., 22: 41-46.

WALLER, R.E. (1982)   Trends in Lung Cancer in London in relation to exposure to diesel fumes. Environ. Internat., 5: 479-483.

Editor's Note

Sir Richard's very careful and in depth assessment of the three
very large and complex IP studies that have been summarised in
the paper delivered by Dr. Alderson, and the appraisal that
Sir Richard has made on their interpretation and inter-relation
will be of lasting value to the field of epidemiology. In the
course of his remarks he has put a question to his audience, in
respect of past smoking habits in the oil industry. To this a
full answer cannot be given, but in view of the interest it
aroused the subject merits comment. The studies did not collect
data as to smoking habits as, retrospectively, this would not have
been feasible. However, it would be an opinion amongst those in
the industry in the period in question, that smoking habits were
not noticeably different because of smoking restrictions within the
plant as it was the practice to provide designated "safe smoking
areas" at appropriate locations which augmented permitted smoking
at mealtime breaks in canteens. A small questionnaire at one
offshore oil platform where there were similar restrictions at
work inferred that if a man bought a packet of cigarettes a day,
he tended to smoke them within the 24 hours even if there were periods
of forced abstention.

The current epidemiology survey "Health Watch", which in his paper
Dr. Leese notes is taking place within the Australian Petroleum
Industry Health Surveillance Programme, has in its 1981
Annual Report tabulated contemporary findings on smoking habits in
the Australian petroleum industry. These are of interest in that
they support the above stated view within the IP, by showing
and on an age standardised basis, that there is a higher ratio
particularly in the older age groups (30 upwards) of those who
smoke compared to the findings in other reported community and
nationwide sampling studies in Australia. Within the oil industry
sample, it is of further interest that employees in certain
categories ("process work", "multiprocess and plant", and "transportatio
and storage") where work restriction limits smoking (as in this
country) form a higher percentage of smokers than do oil industry
office and other workers who are not so restricted (1036 out of
2405, as compared with 521 out of 1553 in unrestricted jobs).*

If a similar smoking prevalence were to have occurred among the UK
Epidemiology Refining and Distribution Study Populations it would
obscure any given explanation for the low death rate from cancer
of the lung.

*   The AIP survey data is reproduced in the next section.

SMOKING PATTERNS

Contribution from the Australian Institute of Petroleum.

In view of the preceding interest in the subject of smoking within
the petroleum industry, gratitude is expressed to the A.I.P. for
the following extract contributed from its Annual Report 1981 on the
Australian Petroleum Industry Health Surveillance Programme
"Health Watch".

" Direct validation of smoking statements would not be justified in
the Health Watch study as such validation can only be done on small
samples and further, Health Watch is not primarily a study of
smoking behaviour. Nevertheless since such data will be used as an
indication of a confounding variable, it is necessary to ensure that
smoking statements are not grossly 'out-of-line' with other
Australian studies.

In Tables 2.4.2 (a) and (b) the reports from men working in the
petroleum industry are compared with those from a community study
in Busselton W.A. in 1972, and from a nation-wide sample reported in
1981. Table 2.4.2 (a) compares percentages of current smokers by
age, in the three studies, and Table 2.4.2 (b) the percentages of
those who have never smoked. Whilst there is variability,
especially in the under 30 age group, there is no particular reason
to suspect biased underreporting. In Figure 1 the proportions
of those who have 'never smoked' is compared, within age groups, with
that reported from the Australian Bureau of Statistics national
survey carried out in 1977. The closeness of the Health Watch
data, on this particular aspect of smoking, to that of the ABS
leads to further confidence in the Health Watch data. Once more there
is less smoking in the youngest groups working in the petroleum
industry; this may well represent a selection into the employed
population as opposed to the general population. It may also be an
effect of pressure within an industry where smoking 'on the job'
is actively discouraged if not forbidden."

TABLE 2.4.2 (a)

Percentages of current smokers, by age, in three Australian studies.

Age	20-29	30-39	40-49	50-59	60-69	
HEALTH WATCH	35	41	40	34	29	38%
BUSSELTON, 1972[1]	38	36	40	36	32	36%
HILL & GRAY, 1981[2]	51	40	43	43	22	41%

Sir Richard has also made an observation to epidemiologists namely that emigration should not be overlooked in their accounting, and this is a most apt comment. In the case of the Refinery and Distribution Centre Studies, the accountability is as follows:

Study	Number eligible in study	Number recorded as "emigrated"	Number untraced	Percentage "untraced"
Refineries	34,781	679	73	0.2%
Distribution Centres	23,358	6	52	0.2%

It is possible, as Sir Richard states, for some persons who have emigrated not to have been identified as having done so, in which they would presumably end up categorised in the "untraced" column. The authors pointed out that none of the untraced were included in the analyses of any part of the studies, and that the overall no-trace rate of 0.2% is extremely small, so that even if all were taken as "deceased" it was unlikely that this would greatly alter the results.

Sir Richard also notes with relevance the possibility, in a retrospective data collection, of records of persons having died being lost. The low "untraced" rate of 0.2% in both these studies indicates this not to have been a problem in the present cases.

TABLE 2.4.2 (b)

Percentages of 'never smoked', by age, in three Australian studies.

Age	20-29	30-39	40-49	50-59	60-69	
HEALTH WATCH	48	30	29	25	19	34%
BUSSELTON, 1972[1]	43	40	32	26	31	34%
HILL & GRAY, 1981[2]	36	39	31	24	27	33%

Smoking habits in men who have ever working in one
of five major job categories.

	Process Work	Multipurpose & Plant-wide	Transport & Storage	Office, non-field	Other
Current cigarette	170(41.5)	449(34.5)	388(45.0)	268(30.2)	230(34.6)
Current pipe/cigar	4(0.1)	17(1.3)	8(0.9)	17(1.9)	6(0.9)
Ex-smoker	101(24.6	355(27.2)	226(26.2)	237(26.7)	163(24.5)
Never smoked	135(32.8)	482(37.0)	240(27.8)	366(41.2)	266(40.0)
	410	1,303	862	888	665

1.   Cullen, K.J., (1975), Alcohol, tobacco and analgesics, Med. J. Aust., 2:211.

2.   Hill, D., Gray, N., (1981), Patterns of tobacco smoking in Australia, Med. J. Aust. (in press).

Fig. 1   Percentage 'never smoked' by age groups in
          Health Watch as compared with Australian
          Bureau of Statistics survey (1977)

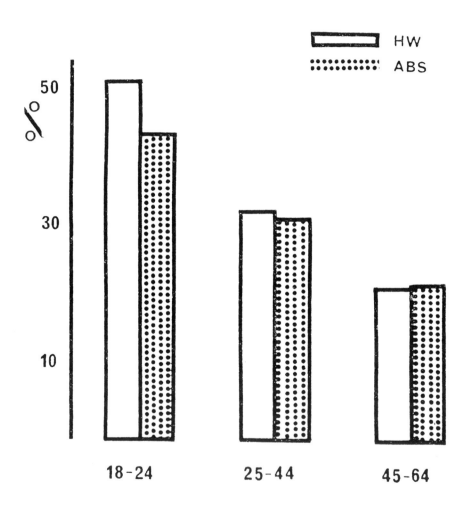

Subsequent written question submitted by <u>Dr. D.M. Davies (BP International)</u> to Dr. Alderson and Sir Richard Doll, in respect of the paper "Epidemiological Studies in the UK Oil Industry".

There is some concern, as you have indicated, that with diseases like leukaemia where the primary cause of death may be something else like pneumonia or other infection, death certification may have leukaemia as a contributory factor or even may miss it altogether. When sampling death certificates, therefore, there is a distinct possibility that some cases of leukaemia may be missed. Have you any 'feel' for the number or proportion of cases missed in this way when carrying out mortality studies on large populations from death certificate information alone?

Dr. Alderson:

Though I have not exact figures with me on leukaemia, I do not think this is a particular problem with this disease. There has been evidence of a rise in this and other countries, and incidence has risen particularly in children. It is not known to what extent this is due to improvement in diagnosis or reporting. As far as our study is concerned, we were comparing causes of deaths reported in Oil Industry workers with mortality in the population; there was no evidence of differential bias in the validity of the certification of our study group.

Sir Richard Doll then wrote on this question.

Sir Richard Doll:

In my experience about one sixth of all adult cases of leukaemia may be missed by counting only those that appear as the underlying cause of death on death certificates. The proportion of childhood cases will, of course, be much higher as so many sufferers are now cured. In one study of 14,000 patients treated for ankylosing spondylitis, in which the development of leukaemia was of particular interest we observed 1759 deaths, 31 of which had leukaemia recorded as the underlying cause. Further investigation revealed 5 additional deaths in which leukaemia was referred to on the certificate as an associated cause (which corresponds closely to Dr. Alderson's experience of 6 cases additional to the 30 deaths attributed to leukaemia) and 2 deaths attributed to aplastic anaemia, which on haematological review seemed more likely to have been due to leukaemia. These last died about 25 years ago and similar errors are less likely to occur now. Most cases recorded as associated causes of death are cases of chronic lymphatic leukaemia (which is not caused by benzene). Acute and myeloid leukaemias are still uniformly fatal diseases and no cases caused by benzene are likely to be missed through not being recorded on death certificates, though an odd case might masquerade as aplastic anaemia.

MEDICAL SUPPORT OFFSHORE

Professor J. Nelson Norman

Professor of Environmental Medicine, Institute
of Environmental and Offshore Medicine,
University of Aberdeen

In any discussion of the provision of medical support for a changing
oil scene the operative expression refers to change, for there
must be few branches of medicine in which the doctors have been
required to adapt and change with such rapidity to follow the
activities of their patients. Initially when the work of the oil
industry was in remote and isolated parts of the world an oil
company merely provided those facilities which were necessary
adequately to care for its personnel and their dependants. This
might have been a small clinic or a major hospital and the local
populous usually benefited greatly because there was nothing else.
Medically speaking what was generally involved was a general
practice of a slightly special type, together with the basic
tenants of occupational medicine referable to the oil industry.

There then occurred the advent of offshore drilling in the Gulfs of
Mexico and Arabia and at once there was a whole host of new concepts
which had to be provided for by the medical advisers to the
industry. The question of time and distance which separated the
doctors from their offshore patients was the first major
consideration to be addressed and this was as important in the
management of minor ailments as in the medical aspects of major
disaster. Then there was the biotechnology of underwater work and
the requirement of keep ahead of growing diving technology in
terms of medical support. The problem of environmental health
assumed new dimensions when consideration was given to the
particular problems of the small, over-crowded, offshore communities
and added to that was the psychological problem of over-crowding and
the psychological aspects of the twenty-four hour work shift and the
associated family problems of separation. These problems were
beginning to sort themselves out when the action moved rapidly to the
support of a very large work force in the ice-bound slopes of
Alaska and it then returned to the oil and gas fields at the
southern sector of the North Sea and finally to the deep waters and
the extremely hostile, remote environment of the northern sector of
the North Sea.

I would like to pause at this point to review the emergence of the
system of medicine which has been developed for the North Sea by

the medical advisers to the oil industry, because I believe it to be an excellent model upon which to develop the newer and even more hazardous areas of the future - the waters off Eastern Canada with their fog and icebergs, the current project in the Beaufort Sea pioneered by Dome, Canada, and eventually the southern oceans and land mass of Antarctica itself. It would be nice if this could be done by a process of accepting what has been achieved by industry in the North Sea and building upon it, rather than by re-inventing the wheel every time a new area is explored.

Let us first of all look at some of the problems which the medical departments of the industry had to face in determining what had to be done in the North Sea. These problems are summarised in Table 1 and revolve around the very large number of operators working in the North Sea. Most oil companies which have been heard of are now represented and a few which have not been heard of, together with an enormous number of contracting companies working from a whole chain of east-coast bases in Britain, and also involving a large number of countries around the North Sea - Norway, Germany, Holland, Belgium, etc. Another main problem was the existence of a highly developed National Health Service in Britain and health services in these other countries also, together with the large number of government departments involved in the various countries and not all of which had similar systems. There was, of course, a very large work force working in extremely hostile conditions associated with the new problems of diving, survival in cold water, etc. The answers were eventually provided by industry itself, in associations with those government departments involved and with the specialist units and inter-disciplinary committees which emerged - but the aim has always been for industry to be one step ahead of interference from outside - and this has meant the gradual establishment of appropriate rules and guidelines which suit both industry and government in matters of health care. In this respect the North Sea industry has much to be grateful for to the succession of excellent Chairmen of the Medical Sub-Committee of U.K.O.O.A., for a very great deal of experience, judgement and hard work was needed. It is interesting to see the emergence at this time of the Eastern Petroleum Operators Association Medical Sub-Committee in Canada, which will hopefully benefit greatly from the work already done by the Medical Sub-Committee of U.K.O.O.A.

One of the big problems which the pioneers of the North Sea had to face was the strength and composition of the health care team which would be necessary and once again many factors had to be considered in the different installations and in the different situations which had to be catered for and these are summarised in Table 2. The time and distance separating the doctor from his patients was the major consideration in this regard, together with the size of the work force and the nature of the work. For example, if a considerable amount of diving or drilling was undertaken then the risk was different from a production platform. The communication

system which in operation is of vital importance in the
determination of medical supervision as the mode and time taken
for evacuation of the sick and, of course, the initial state of
physical fitness of the work force. All these factors made it
obvious that there was a paramount need for flexibility and
versatility if the most appropriate form of health care was to be
provided and it meant that the possible components of the health
care team were reasonably complex. Whatever the system chosen it all
revolved around the person of the company medical officer, who is
normally located offshore and has at his disposal an offshore team
directly under his charge, together with a selection of specialist
services onshore behind him which he has access to if he wishes to
use them, but which are available to a number of companies.

The variety of offshore personnel normally found in the North Sea
installations may be seen in Table 3. Offshore, medical
responsibility may be primarily held by a doctor, but this would
depend on the various factors summarised in Table 2, and it would
be more usual for primary medical responsibility to be held by a
Rig Medic who is normally a nurse specially trained and either
originates from the hospital services, or from military areas.
Most operating companies have now recognised a need for all the
offshore personnel to be trained in appropriate forms of first-aid
and the personnel themselves therefore form the first line of
medical provision, both for themselves and for their colleagues.
The number of Para-medic classifications is not great, however, in
the North Sea, and the second category is the Life Support Escort.
This implies a man of considerable intelligence who has another
job, who has already done a first-aid course, but who is
particularly interested in things medical. He has therefore
undergone a further course of training to allow him to escort the
seriously sick from the offshore installation to hospital and he is
trained is such things as supporting the airways, taking blood
pressure readings, looking after intravenous infusions and communi-
cating medical information. Each installation provides a sufficient
number of such escorts for their requirement; this obviates the
need for a second Rig Medic and also allows a certain amount of
versatility to be built into the safety system in order to provide
the necessary support for the occasional disaster situation which
may occur. The third category of Para-medic found offshore is the
diving Para-medic who is in a particular category since he has special
knowledge to allow him to care for the highly specialised forms of
medical problem which may arise in divers. He is usually a diver
himself and has a fairly extensive knowledge of physiology to begin
with, and upon this the special requirement of both diving and
inter-current illness in saturation systems can be taught. From
this group of personnel are drawn the offshore medical people who
report directly and primarily to the company medical officer.

In Table 4 may be seen a summary of the specialist services which
are available to the company medical adviser if he wishes to use
them, but which are not normally part of the personnel of the

company itself. This includes a group of retained primary care physicians, which is usually present in those centres supporting a substantial degree of offshore activity. There is also a group of diving medical specialists; there are surgical teams which include anaesthetists and which are available for extrication manoeuvres for support in disaster situations. There are environmental health advisers and there is a co-ordinating unit to provide international specialist advice for unusual problems.

It has already been noted that there is paramount need for flexibility if the best care is to be provided for the seriously sick. The highest rate of injury is in drillers, but if we look at the principles behind the management of serious illness in saturation divers, the tenants of the system of medical care for all seriously sick workers offshore emerges and identifies the form of medicine which has become a speciality on its own. The priorities which have been defined in Britain for the care of saturation divers may be seen in Table 5, where the first priority noted is that of training of the offshore personnel, either divers or topside workers in first-aid particularly appropriate to their working situation and this is backed by a system of clear and precise communications to an onshore centre, so that immediate care can be provided and advice given until the third priority comes into operations, and that is the existence of trained doctors willing to go offshore and actually to the saturation system. Following this priority the next in order of importance is the existence of specialist teams of anaesthetists and surgeons and in this diving situation finally the existence of some form of transfer under pressure facility. This priority is placed last, since it will not normally be used in a life-threatening situation, but only after resuscitation and stabilisation of the saturation divers has taken place in order to make his further care more comfortable and more close to expert and laboratory advice.

The medical and para-medical training needs for this work has reached an advanced stage of development, but there is still some way to go if high standards of nursing and medical care are to be maintained within an acceptable career structure for these personnel. It has been clearly recognised for some time that as the work has to be undertaken in more and more difficult circumstances and involves more and more people, that their effective medical care can only be undertaken and the future adequately provided for if there is an element of the academic activity of teaching and research work associated with the development, so that an effective medical support with well-trained doctors can be provided, not only for today, but also for tomorrow when the exploratory activities penetrate the much more difficult areas that we are just entering.

In an area of activity involving several companies there is clearly a need for the emergence of a centre to co-ordinate specialist medical activity and which can be made available to these medical departments of companies wishing to make use of it. This is mainly in view of the need to provide pooled teaching services,

pooled specialist service availability for possible offshore
emergencies, eg. diving, environmental health and hospital based
anaesthetists and surgeons. It is also needed to provide a
research facility to allow for the continuing evaluation of this
new form of medicine and the provision of urgent answers to new
questions when they arise and it is needed so that these services
can be provided with a reasonable degree of practicability and some
economy of resources. The needs are summarised in Table 6, which
emphasises the need for a co-ordinating organisation to act as
the interface between industry, government and medical authorities,
and also for the existence of the communication centre for specialist
services, including diving, surgical and public health services.
It includes also a need for a training centre for oil-related
medicine, for training in first-aid, for the training of Rig Medics,
Para-medics and doctors. It emphasises a need to co-ordinate
appropriate research work and to provide medical liaison between
industry and government on legislation matters, such as public
health procedures offshore, training standards for Rig Medics and
doctors, evaluation of training procedures and of equipment,
together with the identification of necessary research.

In addition to the requirement for such national centres to
undertake this co-ordinating function, there is already a case which
can be made for international co-ordination, since this is a
relatively small community of workers and their managers who move
around the world in an extremely mobile manner. Yet an effective
centre has not emerged in the North Sea as yet - a fact which must
to a certain extent be blamed partly on the laissez-faire attitude
of governments which seem to be satisfied as long as the revenues
appear, because it is interesting to reflect on just how far industry
has had to go alone to achieve the pooled specialist services which
are required, and the contributions which industry has made on its
own are summarised in Table 7, and include the establishment of
public health guidelines and standards of fitness for offshore
workers, the actual provision of onshore hyperbaric chambers in
Britain, and a system of transferring divers at increased
atmospheric pressure, together with the funding necessary to train
doctors in special skills, such as diving medicine, and to provide
specialist surgical teams for the National Health Service and also
the establishment of research work in survival techniques and the
training procedures for helicopter ditching techniques. This is a
formidable array of public services which have been provided by
industry for its own needs and great credit must be reflected upon
the oil industry for pursuing such an intensive course of safe
practice when this has not been required by government
legislation. Considerable time, money, effort, thought and planning
has gone into the system which has now emerged in the North Sea
and this forms an excellent model for the safe practices which will
be required for the exploration of the more difficult areas to be
tackled in the future.

Table 1	The new problems faced by the company medical advisers in determining the system of medicine for the North Sea development.
Table 2	Factors used in determining the extent of medical provision necessary for an offshore installation.
Table 3	The possible components of the offshore medical team on a North Sea installation.
Table 4	The onshore services available to the company medical officer if he wishes to use them.
Table 5	Priorities in the management of serious illness in offshore workers in order of importance.
Table 6	The reasons for existence of a national co-ordinating area for offshore medicine and the uses which it would have.
Table 7	The contribution to pooled specialist services which industry has required to make on its own.

TABLE 1. Problems to be resolved

1. Large number of operators and contractors working from a chain of East Coast Bases.

2. NHS in existence and highly developed.

3. Large number of Government Departments.

4. Several countries involved.

5. Large work-force working in very hostile conditions.

6. New problems - diving, survival in cold water, public health, etc.

TABLE 2. Determination of extent of medical provision

1. Time and distance from medical help.

2. Size of work-force.

3. Nature of work, eg. diving, drilling, etc.

4. Communications.

5. Mode and time of evacuation.

6. State of physical fitness of work-force.

TABLE 3. Provision of medical care offshore

1. A doctor.

2. A rig medic.

3. A para-medic: 1 a) First-aider
                      b) Life support escort
                      c) Diving paramedic

TABLE 4. Onshore medical back-up

1. Retained primary care physicians.

2. Diving medical specialists.

3. Surgical teams.

4. Environmental health advisers.

5. International specialist group.

TABLE 5. Priorities for injury or illness in divers

1. Training in special first-aid.

2. Clear and precise communications.

3. Trained doctors willing to go offshore.

4. Mobile specialist teams.

5. Transfer under pressure facility.

TABLE 6. Co-ordination of medical services

1. Interface between industry, Government and medicine.

2. Communication centre for specialist services:
    - Diving specialists
    - Surgical teams
    - Public health

3. Training centre for oil-related medicine:
    - First-aid
    - Rig medics
    - Para-medics
    - Doctors

4. Co-ordination of appropriate research work.

5. Medical liaison between industry and Government on legislation matters such as:
   - Public health procedures offshore
   - Training standards for rig medics, doctors, etc.
   - Evaluation of training procedures
   - Evaluation of medical and survival equipment
   - Identification of necessary research

TABLE 7. Industrial contributions to offshore medicine

1. Establishment of Public health guidelines.

2. Establishment of standards of fitness for offshore workers.

3. Provision of onshore hyperbaric chamber and TUP.

4. Training of doctors in special skills, eg. diving medicine.

5. Establishment of surgical specialist teams.

6. Establishment of helicopter ditching survival techniques.

HYPOTHERMIA AND SURVIVAL RESEARCH

Surgeon Commander F. St. C. Golden OBE
Royal Navy   -

Senior Medical Officer, Survival Medicine,
Institute of Naval Medicines

INTRODUCTION

In the early exploratory days of the European Offshore Oil and Gas
Industry, the threat to the individual accidentally immersed in the
North Sea was emphasised and clearly understood, and as a
consequence great importance was placed on the requirement for
specialised items of protective clothing and equipment. Most
companies operating in the North Sea have since spared no expense in
providing immersion suits for their personnel in transit to and
from the offshore platforms. But, as pointed out to a similar
conference in 1974 (Golden 1974), the threat is not simply one of
thermal stress; it is a complex problem in which many factors
interact to varying degrees, depending on the prevailing
circumstances. Providing a solution to one aspect of the problem
will not, therefore, necessarily guarantee protection from all.

In the last decade, research has broadened our overall understanding
of this complex problem. This paper reviews the nature of the
threat and in particular discusses the implications of recent
research to our understanding of the problem.

THE NATURE OF THE THREAT

In formulating ones research approach to the problem it was first
necessary to identify the nature of the threat and isolate those
areas where research was required and most likely to be fruitful.

It is now possible to identity four phases where the nature of the
hazard confronting the individual immersed in cold water is different
(Golden & Hervey 1981).

(1)   Initial immersion: the first 2 or 3 minutes. Cardio
       respiratory responses, resulting from the sudden cooling of the
       skin, can produce incapacitation and result in death through
       cardiovascular accident or drowning.

(2)   Short term immersion: from 3 to 15 minutes. The problem in this
       phase is predominantly one of remaining afloat; even competent
       swimmers are unable to swim for relatively short periods in
       cold water (Keatings et al 1969, Golden & Hardcastle 1982).

(3)   Long term immersion: 30 minutes or more. Death at this stage
       is due to hypothermia, or drowning secondary to incapacitation
       caused by hypothermia. Unless a lifejacket is worn and the
       water is relatively calm, death from drowning will occur as
       soon as consciousness is impaired (Golden 1976); this will be
       after about 1 hour in water at 5°C or 2 hours in water at 10°C.
       Thus death is due purely to hypothermia may actually be
       rather uncommon following immersion.

(4)   Post immersion. It would appear that about 20 per cent of
       individuals rescued alive from cold water collapse and die
       during or shortly after rescue (McCance et al 1956).

The importance of the above classification lies in assessing and
planning the preventative measures necessary to reduce the risk.   In
the preventative field of medicine one is primarily concerned with
phases 1 to 3, while in the field of clinical management one is
predominantly concerned with stage 4.

PREVENTATIVE MEASURES

From a preventative viewpoint the most important lesson to be learnt
from the above classification is the recognition of the threat to
life of the immediate and short term effects of immersion. For those
unaccustomed to immersion in cold water, the initial hyperventilation
tachycardia and hypertension make it difficult, if not impossible in
some cases, to perform some vital actions during the initial few
minutes of immersion. These adverse responses can be ameliorated by
habituation or by a reduction in the rate of skin cooling. Thus an
immersion suit which, although not absolutely watertight, will
reduce the rate of ingress of water and thus the rate of skin
cooling, will considerably enhance ones survival chances in this
vital period.   Such a suit, however, will do little to reduce the rate
of onset of hypothermia in the long term, but it is considered that
the greatest threat associated with prolonged immersion in a sea
way is drowning.

Sophisticated, and therefore costly, immersion suits may well delay
the onset of hypothermia but will do little to protect from "wave
splash" with consequent risk of aspiration of sea water. In addition,
such suits are frequently time consuming to don, often difficult to
size, cumbersome to wear and therefore create difficulties in
performing vital actions during or immediately after abandonment.
Alternatively, they do undoubtedly keep the wearer warmer for a
longer period - provided they fit correctly - and thus enhance
survival prospects if immediate rescue is not at hand.

The choice of suit must therefore be based on an assessment of the
risk. Ideally one should protect the individual from the
immediate and short term hazards with some form of suitable
protective clothing while placing the emphasis on long term survival
on some form of rescue or survival craft.

RECENT RESEARCH

The problem of general hypothermia has been well understood for many years and admirably reviewed by Keatinge (1969) and MacLean and Enslie-Smith (1977). Likewise the initial responses to sudden cooling of the peripheral thermoreceptors has been well described by Keatinge et al (1964) and Keatinge and Nadel (1965); however, the association of these responses with the initial problems confronting the individual accidentally immersed in cold water are less well appreciated.

Swimming failure: Recent work by Golden and Hardcastle (1982) has shown the relationship between these acute responses and subsequent swimming failure in cold water. In a group to ten swimmers who had previously successfully swam for ten minutes fully clothed in warm water (Tw25°C), only three managed to swim for ten minutes in cold (Tw6°C) water; the remainder failing at times varying between two and seven minutes. There was no evidence to support either local or general hypothermia as being responsible for swim failure. Exhaustion was also excluded, as oxygen consumption ($\dot{V}_{O_2}$) at swim failure was significantly below (p 0.05) maximum values. There was no evidence of a diving bradycardia at the time of swim failure, in fact the heart rate was significantly higher (p $<$ 0.05) at the time of swim failure than at the corresponding time in the control swim. Both minute volume ($\dot{V}_e$) and respiratory frequency (f) were significantly higher (p $<$ 0.01) at time of swim failure than at the corresponding time in the control swim. It was noticeable that during the first minute of cold immersion, 'f' of the swimmers that subsequently failed increased to 122% (range 56 to 181%) of the control value, whereas the three successful swimmers showed an increase in 'f' of only 35% (range 31 to 37%). Thus it would appear that swimming failure in cold water is associated with sensitivity to cold. The precise cause of the failure is still not certain but it would appear to be associated with an asynchrony between respiration and swimming stroke. Experimental work is proceeding in an effort to determine the precise mechanism.

Post Rescue Death: The other area which received a major research effort in recent years was an attempt to identify the mechanism involved in post rescue collapse and death. The work by Golden and Hervey was recently reviewed (Golden and Hervey 1981) and is still being progressed.

It would appear that the causes of post rescue death may be subdivided into two broad categories as shown in Figure 1.

FIGURE 1. Causes of Post Rescue Collapse and Death.

The problems associated with near drowning have been well reviewed elsewhere (Modell 1971; Conn et al 1980). Those mechanisms associated with the cold/immersion effect are still largely hypothetical and recently reviewed by Golden and Hervey (1981). These authors propose that collapse and death, which occurs during rescue and in the immediate post rescue period, may be accounted for by collapse of the arterial pressure associated with a reduction in the cardiac output consequent to a withdrawal of the hydrostatic squeeze on removal of the victim from the water. In hypothermic immersion victims in whom the work capacity of the heart is much reduced, and in whom the blood volume may be depleted, it seems possible that the sudden withdrawal of the supportive hydrostatic squeeze on the body may be the crucial factor in determining whether or not a viable cardiac output can be sustained. Anecdotal accounts of rescue deaths support the hypothesis of such a mechanism. The only supportive evidence however comes from an unpublished experiment by Golden, Hervey and Slieght (1981) who showed that central venous pressure (c.v.p.) fell by 12 mmHg on lifting a subject vertically by helicopter strop, from cold water. The c.v.p. returned to previous levels when the subject was reimmersed. Ethical problems prevent the pursuance of this experimental approach and animal models are currently being explored.

Rewarming collapse is a phenomenom well understood in the management of geriatric hypothermic patients and reviewed by MacClean and Enslie-Smith (1977). It results from too rapid rewarming of the surface of the body producing peripheral vasodilation and a sudden fall in peripheral resistance before adequate central vasomotor control has returned in a patient who may also be hypovolaemic.

Thus rescue and rewarming techniques may require to be drastically revised but further evidence is necessary before introducing procedures which may in themselves be subsequently proven to be incorrect.

CONCLUSIONS

1. Initial physiological responses to immersion in cold water which may be ameliorated by either habituation of relatively simple forms of protective clothing, have a significant effect on short term survival.

2. Planning for long term survival in cold water should not be based solely on protective clothing; instead the emphasis should be on some form of rescue or survival craft.

3. Many of the problems of post rescue death are clearly understood but further experimental evidence is required to explain some; however, available experimental evidence together with anecdotal evidence supports the hypothesis that collapse of systemic pressure may be a crucial factor in the immediate post rescue period.

ACKNOWLEDGEMENTS

I would like to acknowledge the financial donations received from the following companies, which enable me to purchase the ECG telemetry equipment without which the swimming experiment and some aspects of the research into post collapse would not have been possible:

Amoco; BNOC; BP; Burmah; Chevron; Conoco; Elf; Marathon; Mobil; Phillips and Shell.

My special thanks are due to Dr. R.A.F. Cox who did much of the early "lobbying" on my behalf!

118

REFERENCES

1. CONN, A.W., MONTES, J.E., BARKER, G.A. and EDMONDS, J.F.
    (1980).   Cerebral salvage in near-drowning following
    neurological classification by triage. Canadian Anaesth.
    Soc. J. 27: 201-209.

2. GOLDEN, F. ST. C. (1974).   Hypothermia in swimmers and divers.
    Offshore Medicine Conference, Aberdeen 1974 (unpublished).

3. GOLDEN, F. ST. C. (1976).   Hypothermia: a problem for North
    Sea Industries. J. Soc. Occup. Med. 26: 85-88.

4. GOLDEN, F. ST. C. and HERVEY, G.R.   The 'after-drop' and
    death after rescue from immersion in cold water.
    In: Hypothermia Ashore and Afloat. Chap. 5; pp 37-56.
    Proceedings of the Third International Action for Disasters
    Conference. Ed.J.M. Adam, Aberdeen University Press, Aberdeen
    1981.

5. GOLDEN, F. ST. C. and HARDCASTLE, P.T. (1982).   Swimming
    failure in cold water. J. Physiol. P.C. 17 (in press).

6. KEATINGE, W.R.   Survival in cold water. Blackwells:
    Oxford 1969.

7. KEATINGE, W.R., McILROY, M.B. and GOLDFIEN, A. (1964).
    Cardiovascular responses to ice-cold showers. J. Appl.
    Physiol. 19: 1145-1150.

8. KEATINGE, W.R. and NADEL, J.A. (1965).   Immediate respiratory
    responses to sudden cooling of the skin. J. Appl. Physiol.
    20: 65-69.

9. KEATINGE, W.R., PRYS-ROBERTS, C., COOPER, K.E., HONOUR, A.J.
    and HAIGHT, J. (1969).   Sudden failure of swimming in cold
    water. BMJ 1: 480-483.

10. McCANCE, R.A., UNGLEY, C.C., CROSFILL, J.W. and WIDDOWSON, E.M.
    The hazards to men in ships lost at sea, 1940-44. Spec. Rep.
    Serv. Med. Res. Council No.291; 1956.

11. MacLEAN D. and ENSLIE-SMITH, D.   Accidental Hypothermia.
    Blackwells: Oxford 1977.

12. MODELL, J.H.   The pathophysiology and treatment of drowning
    and near drowning. Charles C. Thomas, Springfields,
    Illinois, 1971.

D. Elliott (Shell UK)

In the protection of an immersed individual against wave-splash
and possible drowning, even when he is provided with positive
buoyancy, the provision of a face guard seems an important
component of any specification for the protective clothing. Would
Surgeon Commander Golden please comment on the possible inclusion
of a snorkel (such as used by the U.S. Navy in their submarine
escape immersion suit), a snorkel which enables the individual
to breathe through the face guard at will?

Surgeon Commander Golden

The snorkel sounds like a good solution to assist the free floating
survivor; however, bearing in mind the location difficulties
in open water, together with the risks of aspiration of vomitus
from sea sickness and peripheral non-freezing cold injury, I consider
that for prolonged survival we should be aiming at some system which
keeps the survivor clear of the surface of the water, eg. one man
dinghies or something similar.

A. Grieve (Shell UK Oil)

The author suggests that early drowning in accidental immersion incidents can be due to a failure of swimming even in known strong swimmers.

Swimming performance becomes weak, rapid and poorly coordinated in association with reflex hyperventilation in response to cold. Some swimmers were not thus affected. Does this indicate that if an individual can exert conscious control over respiration, then swimming would become coordinated and effective, and drowning would be prevented. If this is the case, could teaching and training be given to the workers at risk of sudden cold immersion to prevent early drowning deaths?

Surgeon Commander F. St. C. Golden

Existing experimental evidence suggests that attempts at conscious control of respiration cannot override the initial hyperventilatory responses and thus enable the swimmer to overcome any initial difficulties; it was noticeable, however, in this, and other similar experiments that we have performed, that the really strong swimmers appeared to have sufficient self-confidence in the water to change their normal swimming rhythm, and adjust to the changed respiratory pattern. However, not all good swimmers were able to make this adjustment so I do not consider swimming training is necessarily a practical solution.

The initial sympathetic responses are something which can be habituated against, so that the simple expedient of taking a cold shower may be more beneficial, and simpler, that swim training. However, from a preventive medicine viewpoint, it is much more important to try and safeguard against the worker falling into the water in the first place, by ensuring that adequate safety lines and/or nets are used, while to cater for other specific situations automatic lifejackets and practical protective clothing may be necessary.

DEEP-DIVING AS A LIMITING FACTOR IN
OFFSHORE OPERATIONS

Dr. D.H. Elliott, OBE

Deputy Chief Medical Officer, Shell UK Ltd.

The commercial diver is employed almost exclusively by the offshore
oil and gas industry. For many important tasks the diver is
versatile, effective and relatively cheap, but he has a hazardous
job which is associated with a number of complex environmental
illnesses. With accepted procedures and well-maintained equipment,
accidents are rare. However, as the diver's work extends into
deeper and deeper waters, so the uncertainties increase. Whether
or not the margins of safety diminish at greater depths, it is
apparent that the costs of supporting a safe diving operation will
increase enormously.

The development during recent years of one-atmosphere seabed habitats
and armoured diving suits, such as JIM, and of remotely-controlled
unmanned robots, was at one time expected to eliminate most divers
from offshore exploration and production. It now seems this
prediction, like so many others related to technical innovation, was
premature. Contrary to expectations, the development of so-called
diverless systems is not likely to replace the diver. While the
exploration rig can drill in deeper and deeper waters without a need
for diver back-up, the subsequent phases of field development,
production, inspection and maintenance are still dependent upon human
intervention at depth. Thus, for a number of years to come, the
depth limit for diving will be a depth-limiting factor for offshore
oil and gas development.

To review the health and hazards in the changing scene of deep
diving, an historical approach makes a more interesting start
(Figure 1) than the more conventional classification by such factors
as pressure effects upon the body, the effects of the raised partial
pressures of respiratory gases, the uptake and elimination of dissolved
gases and environmental effects such as cold.

A convenient time to begin a review of commercial diving is around
1960. Up to that time the world record, held by the Royal Navy, was
600 ft (183m) and commercial deep diving was limited to bounce dives
down to some 450 ft (137m), with 30 to 60 minute bottom times. The
development of these procedures has been heavily dependent on naval
research and usually were performed by retired Navy divers using
traditional Navy equipment.

At much shallower depths, led by the U.S. Navy's Sealab and Commandant Cousteau's Conshelf, public attention at that time was turning towards divers who were living at pressure for days at a time in a house on the seabed - a concept that for a number of good reasons has never been accepted by the diving industry.

Against this dominantly naval background of the early 60's came stories of a Swiss mathematician who was making seemingly impossible dives to very great depths and then returning safely to the surface without prolonged decompression stops, in a time that some observers thought should surely kill him. It was said that Hans Keller and his medical adviser Professor Bühlmann were using sequences of "secret" gas mixtures. A 1,000 ft (305m) dive was done under the eyes of the French Navy in their chamber at Toulon and later, with the support of the United States Navy, in the open sea. Keller descended in the bell to a world record of 1,000 ft. There remains some doubt as to just how much he accomplished in the water during his 4 minutes at that depth. This remarkable achievement was overshadowed by the death in the bell of his co-diver during the decompression and the loss of one of their surface support divers.

It has been suggested that the outcome of Keller's 1,000 ft dive subsequently led to two divergent paths of underwater development. Those who witnessed the accident naturally became cautious of sending man to great depths. For instance, the U.S. Navy continued shallow habitat-diving rather than deep excursions or 'bounce diving' and, though possibly co-incidental, the American oil industry followed the diverless route including, for example, the "shirt-sleeve" Lockheed one-atmosphere seabed habitat. The present status and future of one-atmosphere habitats should be covered by the next speaker but, after 15 years or so following the path of one such system, it is perhaps significant that Shell Oil used divers during critical phases of the installation of their Cognac platform in 1977 in 1,025 ft (312m) of water. The 14,000 or more hours spent in saturation by the divers on deep phase of this project is still a record diving achievement.

In contrast to the reluctance in the United States to respond to Keller's record 1,000 ft dive by pursuing a deep diving research programme, some authorities in Europe decided otherwise. They recognised that the Keller dive was a technical break-through, in spite of the tragic deaths associated with it, and that it had potentially important military and commercial applications. The Royal Navy resumed its tradition of deep diving research and Shell in the Hague set up a diving subsidiary with a research programmed based upon Professor Bühlmann's laboratory in Switzerland.

During the 10 to 15 years which followed Keller's demonstration of the potential for deep diving there were many experimental deep dives in naval and unversity laboratories and some sea trials by navies and industry (Table 1). Meanwhile commercial diving developed rapidly in its size and scope while also progressing into deeper waters in support of the oil and gas industry.

The last 20 years must be regarded as a remarkable period in the history of human physiology. The exposure of man in the laboratory to environmental pressures more than 60 times greater than our normal atmosphere at the earth's surface, has been achieved with few setbacks. This has been achieved on a budget which is small compared with that for the physiological component of the space programme where these problems are less profound, and all by a small but dedicated group of researchers and divers in less than some 15 laboratories worldwide. While the limits of human exposure under such carefully controlled conditions have not yet been defined, there are different factors that limit human exposure in the open sea to significantly shallower working depths. In understanding the limits of future diving operations both sets of factors must be reviewed.

BASIC PRINCIPLES

In order to understand the factors limiting deep diving it is necessary to be conversant with partial pressures and with the application of the gas laws to the body. For the purposes of this review, it is sufficient to understand little more than that for every descent through sea-water of 10m (33 ft) the pressure around the diver is increased by 1 atmosphere (1 bar; 100 kPa). Since the tissues of the body are virtually incompressible the direct effect of such an increase of environmental pressure is first noticeable upon the gas-containing spaces. For example, at 10m (2 bar) the gas pressure is double that at the surface (1 bar) and, in accordance with Boyle's Law, its volume halved and at 90m (10 bar) it becomes one-tenth. In order to prevent soft-tissue injury, additional compressed gas must be allowed to enter the gas-containing spaces such as the middle ear and the diver's face-mask.

Secondly, each increase of ambient pressure increases proportionately the partial pressures of the individual gases. Thus the partial pressure of oxygen in air is increased from 0.21 bar at the surface to 0.42 bar at 10m (2 bar) and 2.10 bar at 90m (10 bar). The latter may be referred to by divers as being the "surface equivalent" of 21% oxygen.

While at raised environmental pressure the alveolar gases are being dissolved in the blood tissues of the body at a rate analagous to the uptake of the volatile anaesthetic agents. The mathematical treatment of this dynamic process is complex but it may be considered that the nitrogen or other gas is being absorbed by the various tissues of the body at different exponentially-decreasing rates until equilibrium is reached in each tissue, a process which is complete in some 6 to 8 hours. Once this "saturation" has been achieved, no further gas can be absorbed by the body at that depth. However, if the diver descends deeper, gas absorption would resume, tending towards the new equilibrium.

Thus the effects of raised environmental pressure can be considered as:-

i)   the direct effects of pressure upon the body tissues;

ii)   the indirect effects of the increased partial pressures of the respired gases;

iii)   the uptake of respiratory gases into solution by the tissues of the body and the consequences of bubble formation when the environmental pressure is subsequently reduced.

This is not the place to detail the physiological effects of pressure nor the pathogenesis of the various occupational illnesses of divers but, when considering the limits to deep diving, it is necessary to review these problems in so far as they affect the health and safety of divers.

DIRECT EFFECTS OF PRESSURE

Molecular and Cellular. Most physiological and biochemical processes in the body are affected by alterations of hydrostatic pressure to a greater or lesser extent either because some pathway is associated with a change in the nett total volume of the molecules involved or because, while fluids are indeed virtually incompressible, there is a significant difference in hydrostatic compressibility between the watery and the fatty tissues (Macdonald 1982). At the same time, while the density of water is increased by only 4% at 1,000 bar, its viscosity increases greatly and it shows other seemingly anomalous properties. Gels become liquefied and, for instance, cytoplasmic streaming and cleavage are diminished in tadpoles at 400 bar. There is an increased ionisation of, for instance, magnesium salts in the 200 bar range which has been invoked to account for an inhibition of polyphenylalanine synthesis in animals. The effect of pressure on enzymic reactions is complex and reversible depolymerisation of some proteins occurs. Sodium and potassium flux in unstimulated nerve is increased and the bulk compression of a cell membrane bilayer can indirectly affect the functioning of ion channels and enzymes.

These few examples may not necessarily have any direct application to the relatively shallow diver but they serve to demonstrate the complexity of the physiological effects of pressure on every cell in the body.

Whole Organisms. The behaviour of whole animals under raised hydrostatic pressure has been studied extensively (Macdonald 1982). Pressure will reverse the narcosis of anaesthetised tadpoles. Regnard (1891) showed that pressure alone would immobilise small crustacea. More recently, Kylstra (1982) has reviewed the effects of hydraulic compression upon fluorocarbon-breathing mammals. In the absence of any effects due to the usual respiratory gases, tremors and spasms of the limbs have been demonstrated and, at 125 bar, the respiratory muscles of liquid-breathing mice are paralysed (Lundgren & Ornhagen 1976).

During deep diving trials at the Royal Naval Physiological
Laboratory in 1964 it was noticed that divers had coarse tremors,
dizziness, nausea and some vomiting in dives to 800 ft (243m) with
rapid compression (100 ft min^{-1}; 30m min^{-1}). Many explanations were
offered for these and other phenomena such as myoclonic jerks
and microsleep found in subsequent dives at other laboratories.
Such phenomena are known to be time-dependent, being accentuated
by rapid rates of compression and returning towards normality during
the first 12 or so hours at depth. It is now considered that this
"High Pressure Neurological Syndrome" (HPNS) is a direct effect of
hydrostatic pressure upon neurological function (Bennett 1982).
One hypothesis suggests that the cause of HPNS is a relative
difference in the degrees of tissue compressibility. Cell
membrane structure and thus the transmission of nerve impulses is
disturbed. Some effects of HPNS are responsive to drug or other
agents which themselves affect membrane function. Thus the term
HPNS is a single phase covering many complex alterations of
neuro-physiological function. The possible reversal of HPNS by the
addition of an anaesthetic agent was suggested by the known
reversal of anaesthesia in tadpoles by pressure. The optimum
rates of compression and the relative merits of breathing either
pure oxygen-helium mixtures, or oxygen-helium with some 5 to
10% of nitrogen as a narcotic agent, are the subject of ongoing
research in man.

Gas-containing Spaces.   During descent the reduction of gases in
accordance with Boyle's Law requires that additional compressed gas
enters the gas-containing spaces of the body. Failure to do so
leads to conditions such as alternobaric vertigo which may play a
role in the aetiology of an underwater accident.

At depth the increased density of the respiratory gases leads
to inadequacies of pulmonary function. This takes over from
cardiovascular function as the rate-limiting factor for physical
work at depth. At the depths which we are now considering, the
density of helium is so great that the diver may feel it necessary
to "suck in the gas and blow it out again". It has been shown that
man can perform light work in the sea very briefly at 501m (1,664 ft)
but in one dive at 2,000 ft (610m) in the laboratory an awareness
of the need to breathe interfered with relaxation and sleep. The
respiratory effects are compounded in water by the need for
underwater breathing apparatus (UBA) which imposes an additional
external work load upon breathing. Whatever the position or
attitude of the body, the vertical hydrostatic gradient over it,
through a relatively small pressure difference, also has significant
effects. Furthermore, the limitations upon respiration at deep
depths appear also to have a central component associated with the
HPNS. Thus the whole topic of pulmonary dysfunction in deep diving
should be regarded as fairly complex.

During ascent, any failure to exhale the excess pulmonary gases
which expand in accordance with Boyle's Law can cause pulmonary
barotrauma, "burst lung." Breath-holding by the diver during ascent
is the most obvious cause, but it also follows ascents in which the

venting of excess gas appears to have been satisfactory in persons who have no demonstrable pulmonary pathology. The expanding gases may track through the perivascular tissues to the hilum of the lung, rupture into the pleural cavity or escape into the blood stream causing an arterial gas embolism. Characteristically, following a rapid ascent, the diver loses consciousness immediately. Occasionally onset may be delayed for many minutes. A number of individuals become disorientated and confused, others may have a hemiparesis or monoparesis. A few die almost immediately, apparently from cardiac arrest which may be due to coronary embolism or reflex due to vertebro-basilar embolism.

INDIRECT EFFECTS OF PRESSURE

The increase in partial pressure of the respiratory gases leads to a number of phenomena the effects of which can largely be avoided.

Nitrogen Narcosis. When breathing compressed air or oxy-nitrogen mixtures, nitrogen has a deleterious effect upon mental and physical performance proportionate to its partial pressure. An average decrement of 10% in performing fine manual tasks and with mental arithmetic is found at 50m (6 bar) in a dry chamber. Some persons could have a significant impairment of memory and of decision-making at that depth, a depth at which a doctor may have to attend a diver under treatment for decompression illness. The apparent effects of narcosis may be obscured by the greater concentration of the subject upon the task in hand. The consequent reduction of his ability to be alert towards other events, "perceptual narrowing", needs to be considered in the context of diving accidents. Therefore, at greater depths than 50m, the nitrogen is usually replaced by some other gas as the necessary oxygen diluent.

Oxygen Toxicity.

Pulmonary - The pulmonary toxicity of oxygen is of importance only in dives of relatively long duration. The threshold dosage (CPTD) is ill-defined but, as a guide, it may be said that an oxygen partial pressure of 0.4 bar (40 kPa) may be breathed indefinitely, that some individuals may show toxicity after a few days of breathing 0.6 bar oxygen and that, at pO_2 2.0 bar, pulmonary toxicity will develop in a few hours.

Neurological - Oxygen is also a neurotoxin and, if breathed at rest at pressures greater than 2.8 bar, may cause an epileptiform convulsion. During hard physical work in the water the threshold is lower, approaching 1.6 bar. If underwater, a convulsion can lead to drowning. The latent period to onset shows great individual variation. Periods of rest from high oxygen levels are known to postpone the onset of acute oxygen toxicity and in the dry chamber interludes breathing air or other low pO_2 mixture are routine.

Carbon Dioxide.

The alveolar and arterial tensions of carbon dioxide should remain around normal, independent of pressure when at depth. However, there is a tendency towards an increased P_ACO_2 due both to external factors (such as the design of UBA) and internal factors (such as inadequate alveolar ventilation). The effectiveness of soda-lime and similar CO_2 scrubbers depends upon temperature and humidity as well as the gas pathways through the absorbent. In addition to the direct effects of CO_2 retention upon respiration and mentation, CO_2 aggravates nitrogen narcosis and acute oxygen toxicity.

UPTAKE AND ELIMINATION OF DISSOLVED GAS

Compression Arthralgia.

The ill-defined pains that may occur in one or more joints during the compression phase of a dive are thought to be caused by a transient shift of fluid from articular cartilage to the relatively well-perfused adjacent ends of long bones due to gas-induced osmosis which occurs during the changes of dissolved-gas partial pressures of compression (Bradley & Vorosmarti 1974).

Counter-diffusion Phenomena.

Even in the absence of any change of environmental pressure, a change of respiratory gas mixture at depth can cause an adverse effect. If the respired inert gas is different from that in the gas overlying the skin (as may occur when the diver in the dry chamber is breathing from an independent supply) analogous effects can occur. These two phenomena are both examples of counter-diffusion (D'Aoust & Lambertsen 1982), in which the dynamics of isobaric inert gas uptake and elimination can lead to the formation of bubbles in the tissues. These processes have caused urticaria in the skin at 200 ft (61m) in helium-filled chamber when the subjects change breathing gases, and vertigo in the course of an experimental 1,200 ft (365m) dive.

Acute Decompression Sickness.

The deeper a dive or the longer its duration, the greater is the quantity of gas taken up by the body and the longer the time required for its elimination. The evolution of bubbles in the body can be regarded as due to the decompression of tissues containing an excess of dissolved gas. The consequences range from a transient thrombocytopenia to quadriplegia and death. The calculation of safe decompression tables that largely prevent such consequences is well described elsewhere (Hempleman 1982). The treatment of decompression sickness by recompression is usually fully and rapidly effective but difficulties may arise particularly if treatment has been delayed.

Aseptic Bone Necrosis.

Although the aetiology of "dysbaric osteonecrosis" is commonly thought to be in some way related to the inadequate elimination of dissolved gas, a number of other hypotheses exist. Bone necrosis in divers is particularly associated with saturation diving but on the whole it is only the juxta-articular subchondral lesion that has a potential for causing symptoms (McCallum & Harrison 1982).

OTHER EFFECTS

There is more to the physiology of diving than just the effects of increased environmental pressure. The environment is cold and wet. Immersion of the face causes a reflex bradycardia (Hong 1976). Even the seeming weightlessness of total immersion can have an untoward effect: buoyancy reduces proprioception so that in turbid or dark water with negligible visibility, disorientation is more likely, especially if there is some concurrent alternobaric vertigo.

Thermal Balance.

The coldness of the sea demands that the diver has adequate thermal protection against hypothermia in the form of a wet or dry suit. For dives which are deep or of long duration supplementary body heating is essential. Heat loss is accelerated by the greater thermal conductance and capacity of oxygen-helium than of air, an effect which is enhanced by the increased gas densities at pressure. To maintain the body's thermal store in deep oxy-helium dives, and also to avoid the copious secretions which form an immediate response of the upper respiratory tract to the inhalation of very cold gas, it becomes necessary to heat the inspiratory gases. Even in the dry chamber the comfort zone around $32^{\circ}C$ has a bandwidth of only $1^{\circ}C$.

Communication.

Another effect of increased ambient pressure is upon speech, to the extent that vocal distortion may impair communication. This effect is worse when helium is breathed since the velocity of sound in helium distorts speech even at normal atmospheric pressure.

Atmosphere Control.

When not in the water the saturation diver lives with his colleagues in the isolation of a compression chamber. The control of its atmosphere is of critical importance, for instance the oxygen has to be at a partial pressure between 0.2 and 0.5 bar. In last year's 2,250 ft (686m) dive this meant the maintenance of oxygen content at precisely 0.72%.

Another problem can be caused by the inadvertent introduction of contaminant gases. Some, as may be discussed by the next speaker, may arise by "off-gassing" from the paint and other materials locked into the deck chamber but others may be brought back with divers in the bell. Hydrogen sulphide dissolved in the cold sea water can evolve into the bell at the interface between its warmer atmosphere and the sea; divers may return to the bell carrying traces of toxic materials that they have been using in their work or salvaging. Then there is the special circumstances of welding in the dry atmosphere or a sea-bed welding habitat. Are the effects of, for instance, carbon monoxide or ozone directly proportional to partial pressure, or are they modified at raised environmental pressure perhaps by the increased molecular density of the respiratory gases or by possible interactions with other cellular effects known to occur at pressure?

The necessary warmth and humidity of the chamber provide good conditions for the growth of bacteria, in particular the pseudomonas. A number of saturation dives have had to be aborted because of the pain of otitis externa.

Although there is no space to describe all the physiological and medical phenomena of deep-sea diving, let alone to detail the mechanisms and implications of a selected few conditions of practical significance, the above outline does indicate some of the biomedical problems which need to be considered in relation to the health and safety of the working diver.

DEEP DIVING

The dependence of the offshore oil and gas industry upon divers is well established. At present most drilling operations no longer require diver support but, for some phases of construction and for some repairs, the diver has yet to be replaced. The advent of a diverless robot will certainly reduce the exposure of divers to potential hazard but, if the diverless alternative is not fully reliable, some form of diver back-up may be needed. The armoured one-atmosphere diving suits solve nearly all the biomedical problems of manned intervention, but do not yet have the versatility of the diver in responding to some construction and repair tasks. Such developments are beyond the scope of this paper except that they demonstrate the need to determine how deep divers might be expected to perform useful tasks. In addition to the over-riding demand for safety, another limiting factor in the use of divers at great depths will by the cost. When the diverless alternatives have been perfected, the need for diving operations will probably settle at a relatively shallow maximum depth. In the meanwhile, however, deep diving can be regarded as a depth-limiting factor for offshore development.

Current Research and Development.

Last year a team of 3 men spent more than 11 days in a dry chamber at a pressure equivalent greater than 2,000 ft (610m). During this dive, bicycle exercise at 150-240 watts was performed for 5 minutes at 2,165 ft (660m) and 24 hours was spent at 2,250 ft (686m). A consequence of such a laboratory dive is that it might lead the uninformed into believing that there is now no barrier to achieving a similar depth at sea. Unfortunately this is not so. Not only are there many physiological and medical aspects that deserve further study but a dry laboratory dive does not begin to study the practical problems that would be encountered in a dive to a similar depth at sea.

Even if some lesser depth is selected as a target for the
consistent hard-work that would be needed for a pipeline repair, it
is the interface of the diver with his supportive equipment that
will need most attention. The design of the breathing apparatus
to be used, the provision of supplementary heat to the diver, and
the need for good communications are among the more obvious
requirements. Besides these objectives there are others that are
not biomedical, such as the design and handling of the diving
bell and the diving-support vessel. The only dives made at sea to
depths greater than 1,150 ft (350m) have demonstrated both the
expense of conducting such dives and the need for further
development work. By the nature of their business this is beyond
the resources of the diving contractors. Though originally in the
forefront of deep diving, the navies of the world have now limited
their diving operations to around 1,000 ft (300m). As the prime
user would be the oil and gas industry, an immediate limit to the
depth of offshore operations would seem to be the willingness of the
industry to sponsor such research.

Since 1962, Shell's participation in diving research has been unique
among oil companies and, at relatively low cost, has given us
significant practical advantages. The collaborative "Deep Fjord
Dive Project", now being planned for 1983 - 1984 in Norway, has
an ultimate objective of a saturation dive at sea as deep as
1,480 ft (450m) with hard physical work, so as to demonstrate
the practicality and potential safety of routine working dives to a
lesser depth around 1,150 ft (350m).

TABLE 1   Some significant experiments during the last twenty years
           of diving development

1962	H. Keller with Prof. Bühlmann	1,000 ft (305m) at sea
1965	First commerical saturation dive by Westinghouse on the Smith-Mountain dam	200 ft (61m), 5 days
1970	Royal Naval Physiological Laboratory	1,500 ft (457m), 10 hours, laboratory (dry)
1972	Comex, Marseille	2,001 ft (610m), 80 mins, laboratory (dry)
1975	University of Pennsylvania (Prof. Lambertsen)	1,600 ft (488m), 55 mins, laboratory (wet)
	U.S. Navy, Mark 1 DDS	1,148 ft (350m), at sea
1977	French Navy & Comex	1,508 ft (460m), 10 hrs., at sea
		1,644 ft (501m), 10 mins, at sea
1982	Duke University (Prof. Bennett)	2,250 ft (686m), with more than 11 days deeper than 2,000 ft (610m), laboratory (dry)

FIGURE 1

The rate of progress in diving research illustrated by selected
deep dives over the last century:

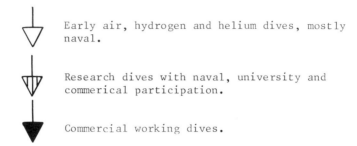

Early air, hydrogen and helium dives, mostly
naval.

Research dives with naval, university and
commerical participation.

Commercial working dives.

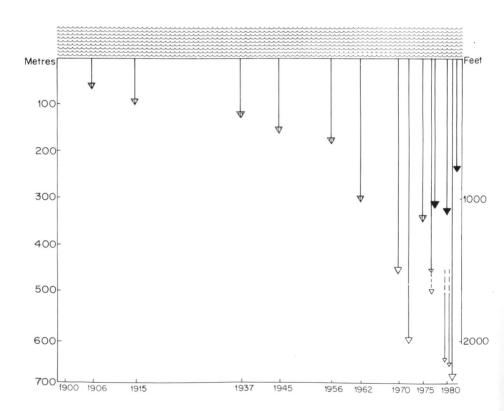

REFERENCES

BENNETT, P.B. (1982)   In "The Physiology and Medicine of Diving",
3rd. Edition. (Eds. P.B. Bennett and D.H. Elliott), Baillière
Tindall, London.

BRADLEY, M.E. & VOROSMARTI, J. (1974)   Undersea Biomed. Res. 1,
151-167.

D'AOUST, B.G. & LAMBERTSEN, C.J. (1982)   In "The Physiology and
Medicine of Diving", 3rd. Edition. (Eds. P.B. Bennett and
D.H. Elliott), Baillière Tindall, London.

HEMPLEMAN, H.V. (1982)   In "The Physiology and Medicine of Diving",
3rd. Edition. (Eds. P.B. Bennett and D.H. Elliott), Baillière
Tindall, London.

HONG (1976)   In "Diving Medicine", (Ed. R.H. Strauss) Grune and
Stratton, New York.

KYLSTRA, J.A. (1982)   In "The Physiology and Medicine of Diving",
3rd. Edition. (Eds. P.B. Bennett and D.H. Elliott), Baillière
Tindall, London.

LUNDGREN, C.E.G. & ORNHAGEN, H.C. (1976)   In "Underwater
Physiology" (Ed. C.J. Lambertsen), FASEB, Bethesda.

MACDONALD, A.G. (1982)   In "The Physiology and Medicine of Diving",
3rd. Edition. (Eds. P.B. Bennett and D.H. Elliott), Baillière
Tindall, London.

McCALLUM, R.I. & HARRISON, J.A.B. (1982)   In "The Physiology and
Medicine of Diving", 3rd. Edition. (Eds. P.B. Bennett and D.H. Elliott),
Baillière Tindall, London.

REGHARD, P. (1981)   "Recherches experimentales sur les conditions
physiques de la vie dans les eaux". Masson, Paris.

HUMAN FACTORS IN MANNED SEABED HABITATS

Dr. D.M. Davies, OBE

Head of Operational Services, BP Group
Occupational Health Centre, British
Petroleum Company plc

(The views expressed in this paper are those of the author and
do not necessarily reflect the policies or intentions of the
British Petroleum Company plc).

SUMMARY

In the present state of knowledge, production of hydrocarbons in
depths of water greater than 500 metres will increasingly have to
rely on diverless techniques. Nevertheless it is likely that human
access to installed seabed engineering systems will continue to be
required, and various techniques are available for this purpose.
This paper is concerned with the manned normobaric dry habitat
approach which enables both systems and men to be housed in
conditions akin to a surface working environment, and discusses the
human factors associated with this approach. It deals with
physiological, psychological and other medical aspects, routine and
emergency life support requirements, and socio-domestic matters, and
concludes that there are no fundamental numan factor barriers to the
use of such habitat systems on the seabed for hydrocarbon production
purposes.

INTRODUCTION

There are a number of methods by which human access to the seabed
mechanical structures associated with hydrocarbon production
from beneath the seabed can be obtained; these are shown in Table 1.
Method No.6, the fixed multi-manned dry normobaric seabed habitat
enclosing the relevant production system structures - the "shirt-
sleeve environment" - becomes more attractive both on operational
and economic grounds as operating depths and surface environmental
hostility increase (Collard and Kemp, 1981). At first sight the
health and safety implications of this system may appear to
militate heavily against its use, but it is hoped that this paper
will demonstrate that there are, in fact, no major health and
safety barriers which cannot be overcome by engineering design and
techniques already available within the current state of the
art.

135

THE CONCEPT

There is nothing new in the concept of men working or living at normal atmospheric (normobaric) pressure in an enclosure surrounded by air or water at much lower or higher pressures - obvious examples being the internally pressurised cabins of high flying aircraft, spacecraft, and the shirt-sleeve environment beneath the sea of submarines. In fact, one can go back over 200 years to find the first effective system - the American one-man submarine TURTLE, which attacked the British frigate EAGLE from under the water in New York Bay. The submarine went through various stages of development during the next 160 years, but remained limited by the need for frequent surfacing to gain access to fresh air to revitalise the internal atmosphere or to enable diesel engines to be run to charge electric batteries. In the early 1940s an important advance was made with the development of the Schnorkel tube by which the submarine could gain access to the surface air whilst remaining fully submerged, but in the late 1950s a more fundamental advance was made with the advent of the nuclear-powered submarine. The generation of power from nuclear sources does not require the use of oxygen, as all fossil-fuelled power sources do, and the concept of the true submarine which could remain totally submerged for weeks or months on end, with no requirement for access to surface air, was finally realised. However, a nuclear-powered submarine could not achieve its full potential unless the crew of up to 150 men could be kept not only alive but physically fit and mentally alert throughout while the submarine remained dived for up to three months; this necessitated the development of sophisticated life-support systems totally independent of access to fresh air. The fact that nuclear submarines have been operational now for approaching thirty years indicates clearly the success of such life-support developments, and it is this long and successful experience of maintaining submarine crews in peak physical and mental condition, and the human factors lessons learnt (see Davies 1973, 1975), which makes the development of normobaric shirt-sleeve environments for use in seabed hydrocarbon production systems so eminently feasible.

The human factors which need to be considered in such developments are listed in Table 2. In order to examine the implications of each of these in a seabed habitat it is best to examine a specific situation, and the concept chosen to be examined is a permanent seabed habitat housing 50 men living and working continuously for two weeks and then exchanging with another crew from the surface on a 2-week shift cycle. The reference design is a cylinder 65 metres long and 12 metres wide, with an internal breathable, ie. floodable, volume of about 5,000 cubic metres, a configuration not far removed from that of a large nuclear submarine. This Habitat, shown in Figures 1, 2 and 3, houses the men in both living and working accommodation, and all the control systems concerned with hydrocarbon production, in a single cylindrical module called the Habitat. This Habitat is attached to a water injection module of identical size also ventilated with fresh air, and to three nitrogen-inerted hydrocarbon processing modules of the same size which are normally unmanned. For purposes of prolonged maintenance, the

nitrogen-inerted modules can be purged with fresh air, but
access normally will be by positive pressure breathing apparatus,
either self-contained or via built-in breathing systems. The
modules are connected by an independent pressure-resistant access
and escape passageway which leads to two independent safe "Havens",
and each of the latter is provided with two independent escape
"Capsules". Normal transport to and from the system is by means
of a dry transfer submersible from a permanently stationed surface
support vessel.

ATMOSPHERIC CONTROL

The basic closed-environment atmosphere control system used in
nuclear submarines is shown in Figure 4, and in the current
concept it is required to provide oxygen at the rate of about 50
litres per man per hour and remove carbon dioxide (CO_2) at about
the same rate. Oxygen can be supplied from bottled HP oxygen, by
the burning of oxygen candles, or by the electrolysis of sea
water obtained from outside the Habitat; the first two methods
are logistically limited and the third is dependent on a plentiful
supply of electrical power. CO_2 is easily removed in chemical
scrubbers or by molecular sieves, but in both cases it must be
collected, compressed and pumped overboard against the ambient sea
pressure. Carbon monoxide (CO) from smoking, cooking, slow
hydrocarbon oxidation, etc., must also be removed from the
atmosphere as it can affect both health and mental performance if
present continuously at levels over about 50 parts per million (ppm)
by volume. Removal is again easily achieved by a CO-burner which
oxidises it to CO_2 to be removed as indicated above. This burner
will also oxidise hydrogen, produced in the electrolytic process
for oxygen production, and from charging batteries, to water, and
will also oxidise most simple hydrocarbon contaminants to CO_2
and water.

It is also necessary to remove dust, aerosols, odours, fumes and
vapours arising from the many activities carried out in the
Habitat, and this can be achieved by the judicious use of mechanical
and active charcoal filters and electrostatic precipitators.
However, the biggest problem in atmosphere control in closed
Habitats is the inevitable and inexorable build-up of what would
be trace contaminants of the atmosphere in an open environment to
unacceptably toxic levels. A list of the more important of these,
their sources, and recommended maximum acceptable levels in the
atmosphere for 14 days continuous exposure (MPC 14 days) are
given in Table 3. Because of the enormous potential range of
atmospheric contaminants concerned - over 800 discrete materials
have been identified in nuclear submarine atmospheres and at least
88 are virtually always present in closed environments (NASA, 1973)
- it is clearly not cost-effective to provide specialised removal
equipment for all potential contaminants, and unacceptable
contamination of the atmosphere would best be prevented by the

establishment of a construction and stores materials control system based on materials of known toxicity. Contamination of the atmosphere by low-toxicity materials is acceptable up to relatively high levels, but contamination of even a few parts per billion (ppb) by some highly toxic materials is totally unacceptable where exposure of the crew to them is continuous and unremitting for the whole of their sojourn in the Habitat. The nuclear submarine is at a great advantage here compared with the seabed Habitat because the former is open to fresh air at the end of each patrol, and the level of contaminants is reduced to zero during these periods, whereas build-up of contaminants in the latter is unceasing and unavoidable during the life of the Habitat on the seabed. Thus the need for contaminant control in the Habitat is paramount, and although the judicious use of materials toxicity information in the selection of construction materials, stores, and equipment permitted on board can ease the problem considerably, it cannot remove it completely. An alternative strategy is clearly necessary, and the fact that the Habitat is stationary on the seabed and must be connected by risers to the surface to discharge the products of its primary activity points to a sound strategy – the use of an umbilical ventilation feed to and from the surface.

Design studies have shown that small-bore ventilation umbilicals can easily provide an air-exchange rate of two Habitat volumes per hour at a supply rate of about 80 m^3 per man per hour at depths down to at least 1,000 metres, and this degree of ventilation provides the immediate answer to the problems outlined above. Because of the large breathable volume of the Habitat and the designed air-exchange rate, the umbilical system will cater for the respiratory requirements of a crew of 50 men with ease, and under normal conditions will prevent the build-up of any contaminants including CO if unlimited smoking is permitted (about 100 litres of CO per day).

Thus no special contaminant removal machinery is required, but because the umbilical may fail and because there could be local build-up of contaminants in areas of poor ventilation, it is still necessary to define MPCs for atmospheric contaminants to which men can be exposed continuously without detriment to their health or wellbeing in the short and long term; examples of 14 day MPCs are given in Table 3.

Facilities for atmosphere monitoring will also be required. The techniques for continuous monitoring for oxygen, CO_2, CO, total hydrocarbon, nitrogen and hydrogen sulphide, by combined or separate instruments are well established and pose no problems. The umbilical exhaust air should also be continuously monitored at the surface for the main parameters as a backup, and for the presence of unsuspected contaminants. Portable instruments should also be available in the Habitat as a further backup to the main system, and so that regular searches for local pocketing of contaminants can be carried out.

THERMAL CONTROL

Not only because the Habitat is thick walled, but also because it is surrounded by sea water at a relatively constant temperature, control of thermal conditions will be easy to achieve and maintain automatically. Experience of one-atmosphere closed environments has shown that dry-bulb temperature and humidity should be maintained in the ranges 20-25°C and 50-70% RH respectively, but that personal control within these ranges should be permitted in sleeping, working and recreation areas.

MICROBIOLOGICAL CONTROL

In fully closed environments there is a marked change in the normal body bacterial populations, gram-negative rods colonising the skin, nose and throat in opposition to the normal gram-positive cocci inhabitants (Morris, 1972). There is a possibility that this change may be detrimental to health, but the continuous introduction of fresh air, and thus of normal gram-positive commensals, via the umbilical would prevent such retrograde colonisation. The possibility of the introduction of pathogenic bacteria at crew change-over remains, however, but this can be reduced considerably by simple medical review of the on-coming crew by questionnaire and medical consultation where necessary.

Bowel organisms could also be spread through the Habitat by venting of sewage tanks inboard but it has been found that even grossly contaminated aerosols are sterilised by circulation of closed-environment air through electrostatic precipitators. An alternative is to vent lavatory areas through bacterial filters, but it is very likely that the umbilical ventilation system will prevent the need either for electrostatic precipitators or bacterial filters, though it will be necessary to vent both lavatory and galley areas through activated charcoal filters to prevent smells reaching the general Habitat atmosphere.

Fungus growth on cold surfaces in poorly ventilated areas is to be expected but can be diminished by anti-fungal paints; it will, however, be a nuisance rather than a health hazard.

FOOD SUPPLY

Good food is most important in terms of crew wellbeing, and pre-prepared airline-type or self-heating canned meals will not be acceptable. Each shift should have its own chef and each "on-crew" period can be made self-supporting in terms of high quality meals by the judicious use of cold and frozen storage, fresh food deliveries at each visit by the submersible, and the installation of proper preparation and cooking facilities.

SANITATION AND DOMESTIC FACILITIES

For a crew of 50, three lavaratories, three showers and one bath will be sufficient to cope with the 25 men who will be coming on or off watch at each change-over. However, urinals will need to be

provided in all working areas in addition, especially in areas
remote from the living area such as the nitrogen inerted
modules even though they are normally unmanned. Deodorisation of
exhaust air from the lavatories will be required as mentioned
earlier, either by passage through charcoal filters or by ducting
into the umbilical exhaust air trunking. Provision for the
storage and disposal of faecal and other waste matter will depend
on IMCO regulations concerning discharges into the sea from fixed
installations, but it is likely that chemical treatment and/or soft-
wall removable containers will be necessary, this and other
waste being transferred to the surface by the transport submersible.

On-board laundering facilities will be required, but if bedding in
the form of nylon sleeping bags and pillows is used, simple
washing and drying facilities for socks, underwear, etc. will
suffice, the bedding being taken to the surface and dry-cleaned
at each crew change.

PSYCHO-SOCIAL FACTORS

Smoking could be unlimited because of the umbilical ventilation
arrangement, but to preserve harmony between smokers and non-
smokers, smoking should be permitted only in specified non-working
areas.

The free availability of alcoholic drinks off-duty will
inevitably lead to both social and management problems, and it
would be best to ban alcohol on board. If this proves
unacceptable, however, only canned drinks should be permitted
and consumption should be strictly controlled by rationing.

The question of permitting free letter and telephone communication
between crew and the mainland is problematical. In Royal Navy
submarines, two systems are used dependent on the nature of the
patrol - either a complete ban or a one-way system from the
mainland to those of the crew who elect beforehand to receive
messages, good or bad. In the current context, if it is acceptable
on receipt of bad news for a man to leave for the surface on the
next available transport, it is good for morale to permit free
communication in both directions. If this is not acceptable,
however, then only out-going mail should be permitted except in
life-or-death situations, as bad news coming in will upset not
only the recipient but also the rest of the crew who will
undoubtedly become aware of the situation.

Crew selection is a subject for a paper on its own but it is unlikely
that personality differences amongst the crew will create
unsurmountable problems. Experience in the Royal Navy has shown that
while certain personality types will be recognised as rejectable
at selection interviews, the majority of applicants for underwater
jobs of this type are self-selecting in psychological terms.

MEDICAL FACTORS

The only medical aspect of importance additional to that of medical fitness for crew selection, and that of the medical questionnaire/ interview prior to going on-crew to avoid introducing infectious disease into the Habitat, is the provision for dealing with medical emergencies. In this respect the situation is little different from that of above-sea platforms in that medical advice and facilities to transfer patients to the surface and then to the shore should be available at short notice. Thus the Habitat requirement is for the provision of first-aid expertise and facilities on-board. If he can be given other employment in the Habitat as well, a rig-medic would provide the best solution for the first requirement, but failing that, all crew members should be given adequate first-aid training to save life under Habitat conditions, and in addition, each shift should contain two fully trained first-aiders.

Regarding material facilities, one single cabin should be nominated and fitted out as an emergency sick bay while still being used primarily for domestic purposes, and suitable transportable stretchers and first-aid equipment should be available in each module. All personnel should be trained to remove casualties from any part of the complex to the emergency cabin after applying life-saving measures. They should also be capable of supporting unconscious casualties on the emergency breathing systems mentioned earlier.

EMERGENCY LIFE SUPPORT AND ESCAPE ARRANGEMENTS

The emergency and escape philosophy for the Habitat under discussion is based on the concept that two basic emergency situations can arise. These are an incident such as fire, plant malfunction, sea water leakage, etc., which renders the Habitat temporarily uninhabitable, and an accident similarly caused which renders the Habitat essentially permanently uninhabitable. In the first case a safe Haven is required where the whole crew can live while the Habitat systems are restored to normality, and in the current design the life support duration is chosen as four days. In the irremediable accident case, a means of escape for the whole crew from the Habitat or from the Haven is required without recourse to the use of the transport submersible. In this case the survival and escape Capsules will be used and each of the four permanently attached Capsules are designed to support up to 25 men for up to 24 hours completely independently of the external situation.

THE HAVENS

The life support system for each Haven will need to be capable of supplying up to 300 cubic metres of oxygen and removing up to 250 cubic metres of CO_2; these requirements will best be met from non-regenerative resources, eg. 150 oxygen candles in two burners, or bottled oxygen, and 1.5 metric tones of soda-lime in canisters to suit the scrubbing system selected. If smoking is to be permitted, a maximum of four cigarettes per man per day will ensure that the ambient atmosphere emergency exposure level (EEL) proposed for CO of 200 ppm will not be exceeded, and precludes the necessity for installed CO-removal equipment. Facilities for the continuous monitoring of oxygen and CO_2 levels must be installed but all other contaminants can be allowed to build up to levels unacceptable in the Habitat because the conditions can be considered as single exposures. EELs for the most likely contaminants are given in Table 4.

The thermal characteristics of the Haven are critical for survival. If sufficient electrical power is available, air conditioning equipment should be capable of maintaining 25°C dry-bulb temperature and 60% RH for the four days. However, in case of equipment failure, thermal insulation must be provided to maintain body core temperatures against a temperature gradient down to 2°C outside the Haven. This will need to be achieved by a combination of Haven boundary insulation, insulated suits, and waste heat from CO_2 scrubbers, self-heating food packs, etc.

Sufficient sleeping bunks for all men must be provided, together with sleeping bags which will aid heat conservation. A half-litre of water and one one-thousand KCalorie packet of sweets per man per day will suffice as emergency rations for four days, but if possible, sufficient self-heating cans of food and drink should be stored to allow each man 2500 KCalories and one litre of each per day; this will aid in maintaining body temperature as well as keeping the men fit for resumption of their normal duties or escape, whichever should prove necessary.

Washing facilities will not be required but chemical toilets and tankage urinary facilities are essential to prevent psychological anuria and constipation in this already stressful situation. An active charcoal filter system will also be required in the Haven to reduce odours and aerosols.

The maintenance of discipline and morale will impose important problems which will be ameliorated by a clear management hierarchy and sound emergency training. A proper balance between work, meal and sleeping times should be maintained with as much routine monitoring and maintenance work as possible, and adequate provision for recreation. If "nerve-cracking" occurs, the leaders and first-aid staff should consider the administration of sedatives or hypnotics as necessary.

Comprehensive first-aid medical facilities are vital as casualties may well be present.

ESCAPE CAPSULES

The essential life-support requirements are an oxygen supply to maintain the ambient level above 137 millimetres of mercury partial pressure and CO_2-scrubbing equipment to maintain the level below 20 mm Hg pp and an automatic pressure-venting system to keep the internal capsule pressure below 1.5 bar. The latter is is necessary because of the small free air volume when fully manned (about 10 cu.m), and the proposed use of a bottled oxygen supply which will be under conditions which approximate to closed-circuit respiration and both oxygen supply and CO_2 removal will need to be regulated carefully. A total of about 20 cu.m of oxygen at 1 bar is required and it must be delivered at a rate of 30 litres per man per hour, while CO_2 needs to be removed at a slightly lower rate. Because the Capsule may not be manned to capacity, either manual or automatic control of these functions is required by reference to continuously-monitored atmospheric levels. Smoking cannot be permitted, though endogenous CO production will not be significant problem. In this life-saving situation, build-up of other contaminants can be ignored, but EELs for 24 hours are also given in Table 4. It is envisaged that Capsules will be released to the surface as soon as it is safe to do so, or after 24 hours in any case, and that the majority of the time will be spent on the sea surface with the Capsule atmosphere open to outside air via a standpipe system in case of rough seas.

Thermal control requirements are as for the Havens, but the potential for hypothermia may be aggravated by exchange of Capsule air for outside air when on the sea surface.

Mobility will be severely restricted but change of position and/or isometric physical exercises should be carried out frequently. Sleeping should also be permitted to conserve atmosphere and energy.

Limited first-aid facilities will be required, but the biggest medical problem is likely to be seasickness on the surface and anti-emetic tablets or self-administered injections will be required. Hyoscine hydrobromide is recommended because the soporific effects of some other anti-emetics may jeopardise eventual successful escape and rescue in some cases; the first dose should be taken by all men on entering the Capsule.

Sanitary facilities may be confined to the bagging of solid excreta. Urination is to be encouraged and the use of absorbent napkins on the body will help to conserve warmth. Food and drink can be limited to emergency rations as described for the Havens.

Life rafts of self-releasing, self-inflating, weather-protected type should be provided in case the rescue services are delayed or unable to locate the Capsules. These should be fitted with all necessary emergency supplies as the escapees may prefer to transfer from the Capsule to them when on the surface. Difficulties in the location of the capsule and/or life rafts by the rescue services should not be under-estimated, and suitable tracking and homing provisions should be made.

CONCLUSION

This paper has dealt with the major human factor problems associated with manned normobaric undersea systems. It is concluded that there are no such problems which cannot be overcome by currently available engineering techniques, and that nuclear submarine and space-transport operations have amply demonstrated the feasibility and practicability of such seabed systems. Thus it is concluded that human factors, per se, should not be considered a bar to the development of working seabed hydrocarbon production installations, and that decisions concerning such developments should be based primarily on engineering and economic considerations.

REFERENCES

COLLARD, M.J. and KEMP, D.A. (1981)   Development of seabed dry one-atmosphere chambers for processing hydrocarbons from deep water, marginal, and sub-ice regions. Paper OTC 3957, 13th. Annual OTC, Houston TX, U.S.A., May 1981.

DAVIES, D.M. (1973)   Sixty days in a submarine: the patho-physiological and metabolic cost. Journal of the Royal College of Physicians of London, 7: 132-144.

DAVIES, D.M. (1975)   Closed environment problems in nuclear submarines. Annals of Occupational Hygiene, 17: 239-244.

MORRIS, J.E.W. (1972)   Microbiology of the submarine environment. Proceedings of the Royal Society of Medicine, 65: 799-800.

NASA (1973)   Bioastronautics Data Book 2ED. Chapter 10. Publication SP-3006, National Aeronautics and Space Administration, Washington DC, U.S.A.

ACKNOWLEDGEMENT

Figures 1, 2 and 3 are reproduced from COLLARD and KEMP (1981) by kind permission of Mr. M.J. Collard of Sir Robert McAlpine & Sons Limited.

TABLE 1.   METHODS FOR HUMAN ACCESS TO THE SEABED

1.   Diving from surface

2.   Diving from hyperbaric underwater facility

3.   Individual in normobaric diving suit

4.   Robots under remote control from surface or
       underwater facility

5.   Manned underwater normobaric mobile vehicle

6.   Multi-manned fixed seabed normobaric habitat

TABLE 2   HUMAN FACTORS IN SEABED NORMOBARIC HABITATS

1. Atmosphere Control

2. Thermal Control

3. Microbiological Control

4. Food Supply

5. Sanitation and Domestic Facilities

6. Psychosocial Factors

7. Medical Factors

8. Emergency Life-support and Escape Arrangements

TABLE 3   MAIN ATMOSPHERIC CONTAMINANTS IN SEABED HABITATS

Substance	Main Source	MPC_{14} days
Benzene	Oils, paints, solvents, etc.	1 ppm (V)
Carbon dioxide	Respiration	3.8 mm Hg
Carbon monoxide	Smoking	25 ppm (V)
Chlorine	Main process, O_2 candles	0.1 ppm (V)
Chloro-fluoro carbons	Refrigerants	*5-500 ppm (V)
Hydrocarbons (total)	Main process	40 mgm/m^3
Hydrocarbons (organic)	Main process	8 mgm/m^3
Hydrogen	Batteries/electolysis	2% (V)
Hydrogen sulphide	Main process	1 ppm (V)
Methane	Processes, metabolism	1% (V)
Methyl Chloro-form	Solvents, adhesives	*2.5 ppm (V)
Nitrogen dioxide	CO burners, sparking	0.5 ppm (V)
Oxygen	(maximum) (minimum)	22% (V) 137 mm Hg
Ozone	Sparking, etc.	0.02 ppm (V)
Sulphur dioxide	Burners, metabolism	1.0 ppm (V)
Toluene	Main process, paints, solvents	20 ppm (V)
Triaryl phosphates	Compressors, oils	0.1 mgm/m^3
Xylene	Main process, oils, solvents	10 ppm (V)

* MPCs set on burner breakdown product toxicity

TABLE 4   EMERGENCY EXPOSURE LIMITS IN SUBSEA HABITATS

Substance	$EEL_{4 days}$	$EEL_{24 hours}$
Benzene	1 ppm (V)	10 ppm (V)
Carbon dioxide	7.6 mm Hg	20 mm Hg
Carbon monoxide	25 ppm (V)	200 ppm (V)
Chlorine	0.5 ppm (V)	2 ppm (V)
Chloro-fluoro carbons	2 x MPC	10 x MPC
Hydrocarbons (total)	10 ppm (V)	500 ppm (V)
Hydrocarbon (organic)	10 ppm (V)	50 ppm (V)
Hydrogen sulphide	2 ppm (V)	5 ppm (V)
Methyl chloroform	50 ppm (V)	250 ppm (V)
Nitrogen dioxide	1 ppm (V)	5 ppm (V)
Oxygen	as for MPC	as for MPC
Ozone	0.05 ppm (V)	0.1 ppm (V)
Sulphur dioxide	5 ppm (V)	5 ppm (V)
Toluene	20 ppm (V)	100 ppm (V)
Triaryl phosphates	1 mgm/m^3	10 mgm/m^3
Xylene	10 ppm (V)	100 ppm (V)

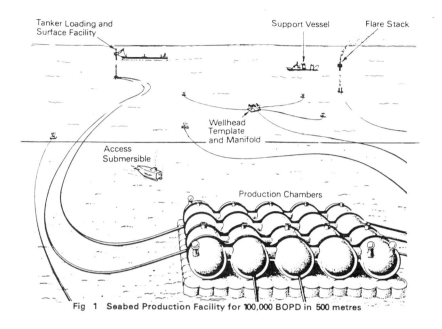

Tanker Loading and
Surface Facility

Support Vessel

Flare Stack

Wellhead
Template
and Manifold

Access
Submersible

Production Chambers

Fig 1   Seabed Production Facility for 100,000 BOPD in 500 metres

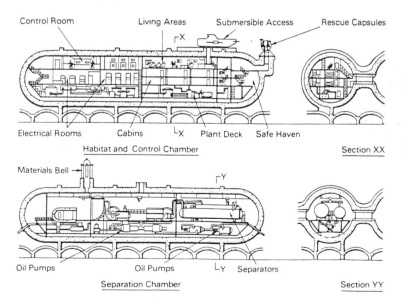

Control Room

Living Areas

Submersible Access

Rescue Capsules

Electrical Rooms   Cabins   Plant Deck   Safe Haven

Habitat and Control Chamber

Section XX

Materials Bell

Oil Pumps   Oil Pumps   Separators

Separation Chamber

Section YY

Fig 2   Typical Sections of Production Chambers

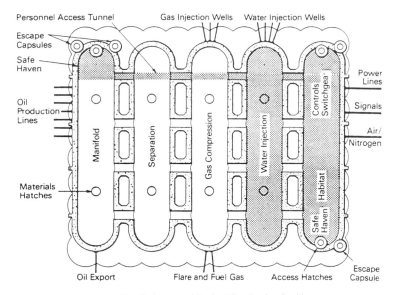

Fig 3   Layout of Seabed Production Facility

Note: — Shaded Areas are Ventilated with Fresh Air Other Areas are Vented with Nitrogen

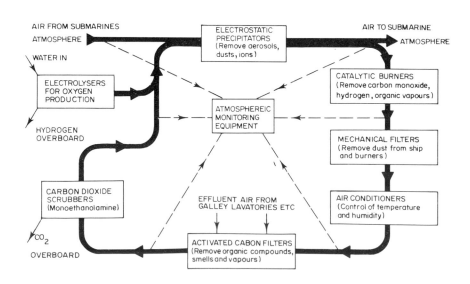

Fig 4 Atmosphere control system for nuclear submarines

NO$_x$ EMISSIONS AND HUMAN HEALTH

H.I. Fuller

Consultant

This presentation is an abbreviated version of the detailed
appraisal prepared by Mr. Fuller on behalf of the Institute of
Petroleum, June 1982 (1).

Aware that in occupational mishaps very high concentrations of
nitrogen dioxide can seriously or even fatally damage the lungs,
environmental authorities seek to know if ambient concentrations of
nitrogen oxides, emitted during combustion, present a hazard to
health. The EEC Commission have suggested limits for nitrogen
oxides (NO$_x$) in urban air. In 1977 the World Health Organisation set
out health criteria for NO$_x$ and proposed exposure limits. The US
Environmental Protection Agency (EPA) has proposed revisions to the
1971 US Air Quality Criteria for Nitrogen Oxides, with a view to
establishing whether a short-term exposure standard is necessary.
These provoked critical reviews, both of the revision and of the
topic itself. A report has been prepared for the Institute of
Petroleum to appraise this situation (1).

Although there are other oxides of nitrogen, and nitric oxide (NO)
is the principal one emitted in combustion gases, only nitrogen
dioxide (NO$_2$) need to be considered for health risk at ambient
concentrations. These concentrations of NO$_2$ are highest in urban
areas, reaching annual averages of 0.03 - 0.05 ppm (60-100 μg/m^3).
One hour peaks are typically 0.1 - 0.2 ppm (200-400 μg/m^3) in cities
like London - Table 1 - but may exceed 0.4 ppm (750 μg/m^3) in the
densest urban areas and climatological conditions such as Los
Angeles or Tokyo. In homes with gas stoves and inadequate
ventilation, 1 hour peaks of NO$_2$ can reach 0.5 ppm (900 μg/m^3),
while tobacco smoke typically can contain 80-110 ppm NO and 10-50 ppm
NO$_2$. In occupational accidental exposures, concentrations can
greatly exceed 100 ppm - Table 2.

Three types of study have been used to assess the health risk from
exposure to ambient levels of NO$_2$: epidemiological, clinical
laboratory exposures, and animal exposures. Each has demonstrated
marked limitations - Table 3.

Epidemiological studies propose to measure responses of real populations to different levels of nitrogen oxides. This is intrinsically a sound approach but the responses must be statistically significant and not due to chance alone. All other factors likely to produce similar health responses must also be measured and their effects statistically eliminated. Monitoring techniques must be specific and sensitive. The 1966-72 US 'CHESS' studies of respiratory illness at Chattanooga and in New York, health studies in Japan, and various other epidemiological studies, have all proved inadequate in these respects. In their extensive 1980 critical review of the NO_x environmental health situation, EPA conclude that epidemiological studies conducted prior to 1973 are of questionable validity, due to difficulties inherent in the analytical technique, that other pollutants (especially particulate materials) had confused any Chattanooga and the CHESS findings and wholly confounded the Japanese studies, and that albeit they had formed the major basis of the earlier standard setting procedure, the assessments of health effects had been statistically dubious. Table 4. Summarising, EPA state that the epidemiological studies "failed to establish an association between disease prevalence in populations and the concentrations of NO_2 to which these populations were exposed".

Studies of whether NO_x from domestic gas stoves in the home leads to increased respiratory illness, especially in children living there, also show confusing results. Table 5. Studies in some gas stove homes with inadequate ventilation giving peaks of NO_2 showed that the incidence of respiratory illness was the same as in electric-stove homes. In other studies, a difference in incidence was detected but EPA acknowledged that other factors such as low temperatures, humidity or parental smoking may have been responsible. EPA claim based upon parental recollection cited in one such study an adverse effect on the health of very young children is produced by indoor peaks of NO_2 at around 0.5 ppm but critics say that 'adverse' studies were inadequate in too many respects to allow such a conclusion to be reached - a point recorded in the recent EPA documentation.

Exposure of human subjects to NO_2 in the laboratory, necessarily limited in both time and severity but ranging well above peak ambient levels, indicates that mild and reversible physical effects on breathing resistance are produced by concentrations of 1.6 - 5.0 ppm; more profound changes set in only at higher concentrations. Most authors find no significant difference of response between healthy adults and chronic bronchitics or asthmatics at the lower exposures. However, EPA perhaps because of their now reclassification of the evidence of earlier key studies as negative or at best uncertain, fall back on one unusual study (Orehek) which implied that for some (and only some) asthmatics, NO_2 exposure made them more susceptible to an aerosol of carbachol - a potent chemical which causes artificial constriction of the bronchi. Both EPA and critics agree that both the methodology and meaningfulness of the results need much consideration and further work to see if the reported findings can be replicated - yet EPA continue to cite the study, in the context of the uncertainty of 'margins of safety'.*

Animal exposures have clarified the qualitative mode of action of NO_2, though mainly at much higher concentrations than ambient levels. Mild and reversible changes in the lung function of mice - a most sensitive species are produced by exposure to 0.2 ppm NO_2, while other species of animal are unaffected by 1 ppm. The susceptibility of mice to deliberate and lethal lung infection with aerosol pathogens is increased by prior exposure to more than 2 ppm NO_2 for some hours. Although this finding supports the unrealised rationale behind the epidemiological work, the levels of infection concerned are vastly different. EPA summarise the animal exposure studies by stating that the lowest dose at which either biochemical change or a reduction in the resistance to infection can be detected in experimental animals is 0.5 ppm over 4 hours or more. All writers including the EPA agree that the results of animal studies cannot quantitatively be extrapolated from one animal species to another - or to man - Table 7.

This appraisal indicates that nitrogen oxides - specifically NO_2 - cannot be a major factor for adverse health effects in the urban environment. If they were, the effects would be more clearly evident. The studies suggest that human exposures of some hours, possibly repeated, to at least 1½ ppm are needed to provoke even a mild and reversible response. Such exposures do not seem to occur even in enclosed environments. There is no evidence that an Air Quality Standard based on average annual values is justified.

(1) Reference: ASSESSMENTS OF THE HEALTH HAZARD FROM NITROGEN OXIDES.

An appraisal by Howard I. Fuller, MA., F.Inst.Pet., for the Institute of Petroleum, June 1982, 56pps. (Available at the Institute of Petroleum).

* The subsequent careful work of Hackney et al (1981), now reported, has shown the Orehek study with a bronchoconstrictor not to be replicated.

TYPICAL AMBIENT CONCENTRATIONS, ppm

LOCATION	NO		NO_2	
	year average	peak 1 hr	year average	peak 1 hr
city streets	0.04 - 0.1	0.2 - 1	0.03 - 0.05	0.1 - 0.2 *
small towns) rural areas)	< 0.03	< 0.15	< 0.02	< 0.05
US Air Quality Standard (1971)			0.05 max	—

* may exceed 0.4 ppm in the densest urban areas and climatological conditions such as Los Angeles and Tokyo.

2

INDOOR CONCENTRATIONS OF NO_2

GAS COOKERS IN ILL-VENTILATED KITCHENS

Peak levels near stove   up to 0.5 ppm

Kitchen generally (while stove alight)   0.05 - 0.15 ppm

TOBACCO SMOKE   10 - 50 ppm

OCCUPATIONAL ACCIDENTS   100+ ppm

3

STUDIES OF NO_2 HEALTH RESPONSES

ANIMAL EXPOSURE

- controlled but unreal environment
- less limit on subjects and exposures
- indicates sites and mechanisms of responses
- results do not extrapolate directly to man

CLINICAL EXPOSURE

- controlled but unreal environment
- limited subjects, short exposures, usually low severity
- evaluates short-term effects of low levels

EPIDEMIOLOGY

- real populations in real environments
- evaluates long-term unspecific low-level exposures
- interferences must be recognised and statistically controlled

4

EPIDEMIOLOGICAL STUDIES

CHESS/EPA AT CHATTANOOGA, 1966-1972

Health records of children in defined areas around a nitric acid plant

Limitations:

- pollution levels in the study areas not distinctive
- unreliable air-quality monitoring
- other pollutants present (eg. ozone, acid mist particulates, etc.) but not monitored
- emissions decreased during studies
- health responses poorly substantiated and/or retrospective based on recollection only

Results:

Inconclusive, no clear effects seen at peaks up to 0.5 ppm NO_2;

EPA (1980) now stress that "other pollutants, notably ozone, sulphur dioxide, sulphates, nitrates were also present in all exposures and especially at the higher NO_2 levels" and that "this mixture, associated with a prolific point source may have been quite different from that often found in urban areas."

5

EPIDEMIOLOGICAL STUDIES

<u>HOMES WITH GAS STOVES</u>

UK STUDIES

- higher incidence of respiratory illness in (urban) children weakly associated with (weekly) NO_2 average level (Melia et al 1977, and 1979)

- other confounding environmental factors maybe not eliminated (family smoking habits; high humidity, linked with low bedroom temperature resulting from ventilation to control condensation etc.)

US STUDIES

- some show no positive effects on respiratory health (Mitchell et al 1974, EPA 1976, Keller et al 1979)

- positive findings involved questionable methodology

6

RESULTS OF CLINICAL EXPOSURES

HEALTHY ADULTS, EXERCISING:

2.5 ppm/ 2-3 hr   Increased airway resistance - short term

1 ppm / some hours   No overt symptoms, even subjective; some inconclusive blood chemistry changes

CHRONIC RESPIRATORY DISEASED SUBJECTS:

1.5 ppm/ 1 hr   No increase in airway resistance

0.5 ppm/ 2 hr   Only normal daily variations in asthmatics and chronic bronchitics

7

LIMITATIONS OF ANIMAL STUDIES

- generally NO$_2$ alone -
  very few studies with 'real' mixtures

- high concentrations for rapid and significant responses -
  few studies at ambient levels

- mostly mice and guinea pigs -
  but wide species variation observed

SUMMARY RESULTS OF ANIMAL STUDIES

0.3 - 0.5 ppm/ several days	) some species show subtle cell
2 ppm/ some hours	) and enzyme changes
1.5 ppm/ 2 weeks	) reduced resistance (of mice) to
2+ ppm/ 3 hours	) severe experimental infection
1.0 ppm/ 4½ years	- no apparent response (in dogs)

8

CONCLUSIONS

– other oxides of nitrogen (eg. NO) are present in the atmosphere, but only NO_2 of potential concern at ambient concentrations

– animal studies identify response mechanisms but are not extrapolatable between species or to man

– short-period acute response is more significant than chronic exposure

– clinical exposures suggest no adverse response at $< 1\frac{1}{2}$ ppm

– no convincing evidence bronchitics/asthmatics more susceptible

– epidemiological studies indicate any effects must be weak

PUBLIC INQUIRIES AND HIGH TECHNOLOGY
PROJECTS

A.C. Barrell

Health and Safety Executive

INTRODUCTION

Large scale projects such as nuclear power or liquefied energy gas,
involving new or complicated technology, can yield significant
benefits to the community. But in some cases they contain the
potential, even though remote, for very serious accidents which
could affect people living in the immediate vicinity. Thus, a
planning proposal to locate one of these facilities on a particular
site is likely to arouse considerable debate.

The dilemma is, that while society as a whole may stand to gain,
individuals living near the proposed site may suffer loss of
amenity and be exposed to a new hazard. Views as to whether the
benefits are worth any extra risks from the new technology are likely
to vary. People who are not immediately affected by the location
of the project may nevertheless feel concern. The controversy, if
sufficiently widespread and sustained, can lead to the setting up
of a public inquiry into the planning proposal.

At such inquiries it is usual for a mass of scientific and technical
evidence to be presented, much of it incomprehensible to the layman.
A common feature is that there is insufficient data to provide
conclusive statistical evidence on the likely performance of the
technology and the chance of a serious accident. Each of the
interested parties may thus provide different estimates of the
probability and consequences of various events. There are no
objective criteria for settling these differences.

It is important that the decision made as a result of the inquiry
is not only justified on the evidence available, but that it and the
whole process of the inquiry is convincing and reassuring to
uncommitted members of the public, particularly those who live
nearby.

The use of tribunals and inquiries as instruments of public
administration is of course nothing new. It has long been recognised
that the balance between private right and public advantage, needs in
many cases to be judged in a system less formal than the courts,
particularly when the issues depend less on the law than on the

163

application of policy. Thus, as the State's authority over the individual has increased, procedures have developed which aim not only at the efficient administration of policy, by speedily resolving disputes and appeals, but also at satisfying the public generally that decisions have been properly made.

HISTORICAL BACKGROUND

The history of these administrative tribunals is a long one. The appointment of local dignitaries to oversee the collection of taxes can be traced back to the fourteenth century, and in Tudor times the Star Chamber (despite its reputation) functioned as an arbiter of disputes separate from the courts and more readily accessible. The Industrial Revolution, and particularly the growth of the railways, brought with them the need for special tribunals to judge the detailed technical wrangles over 'undue preference' and monopolisation; the best example being the Board of Railway Commissioners, which survives today is the Transport Commission.

The modern lay tribunal has its origins in the 1911 National Insurance Act, which provided for appeals to be made against decisions by the insurance officers administering the Act whether or not to pay unemployment benefit (of 7 shillings per week). In 1932 the Donoughmore Committee into Ministers' powers was appointed to examine these quasi-judicial processes, but its terms of reference were something of a straitjacket and the Committee's report (1) was disappointingly limited.

Since the last war, the number of tribunals has increased significantly and it is possible to identify two main sorts: those dubbed by Lord Reading "the camp-followers of the Welfare State" (2), which arose from the social security programme of the post-war Government, and those which have arisen from policies of increasing regulation. Questions of planning and land use clearly fall into the second group, the group which is principally of interest in the context of this paper.

PUBLIC LOCAL INQUIRIES AND PLANNING

The Public Health Act of 1879 saw one of the first attempts to provide for a local inquiry system, married to the controls which it exercised over new buildings with the aim of improving public sanitation. The first town planning legislation in 1909 brought with it a Planning Inspectorate, and the Town and Country Planning Act of 1932 contained the now familiar provision that an applicant who was aggrieved by a refusal of the local planning authority could appeal and the Minister was required, if either the appellant or the authority so desired, to afford an opportunity for them to appear before and be heard by a person appointed by the Minister.

By the early 1950's disquiet had grown (3) about the balance between the rights and interests of public authorities and individuals, particularly in relation to compulsory purchase orders. At this

time, the function of planning inquiries in collecting and assessing
facts and views for the Minister to arrive at his decision, was only
part of the administrative process. The Minister was free to take
policy considerations into account, to consult colleagues and to
take further advice. Planning authorities did not disclose their
case before the inquiry: Inspectors' reports were not published;
their recommendations were not disclosed; and the reasons given for
decisions were frequently very brief.

THE FRANKS COMMITTEE

These causes of dissatisfaction led to the appointment of the
Committee on Administrative Tribunals and Inquiries under the
chairmanship of Sir Oliver Franks, afterwards Lord Franks, in 1955.
Their review was through (4) and the guiding principles which they
thought should govern inquiry procedures - openness, fairness and
impartiality - have been accepted by successive governments. Many
of Frank's detailed recommendations were also accepted and they have
led to the establishment of a system for inquiries which has five
main requirements:

(i)   the planning authority must provide a statement of their
          case at least 28 days before the inquiry;

(ii)   a government department must provide evidence, if it has
          objected to the application;

(iii)   the Secretary of State must give the parties an
          opportunity to make further representations if disagreement
          on a finding of fact or receipt of new evidence inclines
          him to disagree with the Inspector's recommendation;

(iv)   the Secretary of State must give reasons for his decision;
          and

(v)   the Inspector's report must be made available to the
          parties.

Another recommendation of the Franks Committee led to the
appointment of the Council on Tribunals, charged with keeping the
constitution and working of tribunals under review. It reports
directly to the Lord Chancellor, who may, after consulting with the
Council, make statutory rules of procedure for inquiries. Currently
these powers are governed by the Tribunals and Inquiries Act 1971.

RULES OF PROCEDURE

Rules for planning appeals, embodying the five requirements already
outlined, were first made in 1962. The new arrangements worked well
enough to make it possible to transfer certain appeals to Inspectors
for decision, and there are this currently two sets of Rules of
Procedure: the Town and Country Planning (Inquiries Procedure)
Rules 1974 (5) (Secretary of State's Rules) and the Town and
Country Planning (Determination by Appointed Persons) (Inquiries
Procedure) Rules, 1974 (6) (Inspectors' Rules). These Rules apply
to most of the common sorts of public inquiries, but not to all.

Even where the Rules do not apply, successive Governments have undertaken to have all inquiries conducted in the spirit of the Rules.

In major planning inquiries, whether the Rules apply or not, it is becoming more common practice to encourage the principal parties to meet before the inquiry to agree basic facts, to identify the germane issues, and to arrange a practical programme of business for the inquiry. At the inquiry itself, the Inspector has a wide discretion to hear anyone with something relevant to say, and to permit the calling of evidence and cross-examination.

It is fundamental part of the inquiry system that Inspectors may only come to their recommendations to the Secretary of State (or their own decisions) on the basis of the evidence presented to them at the inquiries and supported by information gained at their site visits. The Inspector may not ordinarily seek further information or advice once the inquiry has been closed. If it becomes necessary to obtain further information, the parties must be given an opportunity of considering it.

THE SECREATRY OF STATE'S DECISION

The Secretary of State is not bound by the Inspector's recommendation, although he does follow it in over 95% of cases. If he in his turn takes fresh expert advice or takes into account other new evidence, the parties must be given the opportunity of making representations on the new matters, and the inquiry may be re-opened to test them.

Once the Secretary of State has set out his reasons for the decision in his decision letter, this constitutes the authoritative pronouncement on the application or appeal. He cannot review the decision and its merits cannot be challenged, except that the decision can be referred to the High Court if a question of law is concerned.

In summary, therefore, the effect of Franks has been to change the function of the inquiry to make it central to the decision rather than just a part of the administrative process. The making of Statutory Rules of Procedure has brought greater certainty and fairness to the system, but with these benefits have come a greater formality and some reduction in flexibility. There will always be the conflicting demands of thoroughness and speed.

MAJOR INQUIRIES

Current problems. In planning cases of widespread public interest, there has of late been a tendency to question whether the need for the development has been properly established. This tendency has been most evident in proposals for new motorways, airports, nuclear energy projects, and latterly major non-nuclear industrial hazards.

It is argued that these projects have implications which call for a thorough assessment of the balance between national economic considerations and the effect of the development upon environment, the quality of living, and safety. Such projects are unlikely to be popular locally and, moreover, are promoted by large organisations equipped with expertise and resources far greater than those of most of the potential objectors to them.

To my mind there is no doubt these wider issues can be successfully encompassed within the public inquiry system, notably, for example, at the Windscale Inquiry, showing that the present system is capable of expansion to deal with matters which, being beyond purely local effect, would not normally have featured at an inquiry. Nevertheless, concern and criticism continues to be expressed. This can be seen as part of the public debate about technological change, and the size of the investment that tends to be associated with that change. While suspicion of technological change is nothing new, its current expression has become politicised in the sense that there are now formal movements dedicated to the examination of investment programmes, new departures in technology and so on. Major inquiries into what, in the public eye at least, are high technology projects are increasingly becoming significant in terms of their degree of politicisation. Thus a procedure designed for settling local planning matters is being broadened to encompass the technology-questioning movement. This most often takes the form of a discussion about the 'need' for the particular project.

Establishing Need. All too often when a project goes to inquiry there has been no prior public debate on need, even though the policy underlying the project may well have been developed over many years of deliberation within government. In some cases policy will have been debated in Parliament, but such debates do not always provide an opportunity for a close testing of the case for a particular policy or of the merits and demerits of alternative policies.

A working party set up by the Royal Town Planning Institute (RTPI) has been discussing this problem and considers that the need for a major project and the policies, usually national ones, upon which that need rests, should generally be the subject of analysis, debate and decision prior to inquiry; and that the public should have access to that process. It offers various possibilities (7), including a full-scale parliamentary debate on each project, or Standing Commission of Inquiry reporting to the responsible Minister with the findings being debated in Parliament, or referral to a Parliamentary Select Committee. On the face of it, all these proposals are likely to run up against the obstacles of shortage of Parliamentary time and the limited opportunities they offer for examining witnesses in depth.

The RTPI working party made two other interesting recommendations.
Firstly that alternative locations for, and the detailed siting and
design of, a major public project should be considered by a Project
Assessment Commission, set up as appropriate by the Secretary of
State. The PAC should be responsible for the preparation, largely
by the promoter, of Environmental Impact Statements, summarising
the project's potential impact at each location. Secondly that a
trust should be set up to disburse money to participants at
PAC's and other inquiries on a selective basis.

Advantages. For my part, whilst recognising that important questions
have been raised about the adequacy of the public inquiry process
for major projects, I am convinced that it is still a valuable
institution. It is notable how often some past, present or
proposed course of conduct is sought to be made the subject of a
public local inquiry. It is common ground that inquiries into
major projects, offer an important opportunity for the public not
only to participate in discussion and argument on major individual
decisions, but also at the same time to participate in the analysis
and criticism of the programmes and policies which underly a
particular project. One of the reasons why these opportunities are
important is that the adversarial procedure at an inquiry allows
full use of the instrument of cross-examination. This enables
information to be obtained from promoters, and their arguments to
be placed under rigorous test. When properly informed and
disciplined it can be said to be the most effective instrument
for testing a proposal that is available to the public.

The inquiry is also an excellent opportunity for the involvement
of the individual. The individual's involvement may not often give
rise to new information, but it will usefully emphasise what are
the matters of public concern. Finally, the inquiry is an
opportunity for a promoter to explain his proposals and to meet
and satisfy points which were either unknown to him or the force
of which he had insufficiently appreciated. Much opposition arises
from forgivable but unfounded alarm. The dispersion of that alarm
and the meeting of points are valuable things of themselves.

There is one benefit in a public inquiry which does not appear to be
mentioned in the considerable literature on the subject. This
benefit accrues to the participants themselves. My direct
experience of major inquiries is limited, but I have prepared
evidence for the Health and Safety Executive, and have been
cross-examined at three such inquiries. The discipline of
preparing evidence knowing that one is to be cross-examined, and
then undergoing the cross-examination itself, causes one to
examine policies critically and to become more aware of their
strengths and weaknesses. It is furthermore a valuable exercise
in translating complicated scientific and technical matters into
lay terms. It certainly makes one appreciate more clearly the
impact of one's day to day work on the public at large. I suspect
that this discipline is beneficial in similar ways to the
participants in most inquiries.

TWO EXAMPLES

Two recent examples of public inquiries in the field of chemicals and petrochemicals, namely those at Moss Morran and Canvey Island, are possible pointers to what may be expected in the future.

Moss Morran. Early in 1977, as part of the Flags Project for exploiting certain North Sea gas fields, Shell and Esso applied for planning permission for a natural gas liquids separation plant and ethane cracker at Moss Morran in Fife and an associated marine terminal nearby at Braefoot Bay. Considerable concern was expressed locally by objections to this major project and the matter was eventually called in for public inquiry. The Aberdour and Dalgety Bay Action Group, a well informed and articulate group of local residents, voiced their objections both at the inquiry and subsequently through the media. They claimed, probably with some justification, that their efforts were hampered by the lack of funds to employ necessary legal and scientific assistance.

Nevertheless when the Secretary of State for Scotland granted planning permission in August 1979 he attached a number of conditions. Condition No 24 of the Shell planning consent reads as follows:

"A full independent hazard and operability audit in relation to the design and construction of the NGL feedline within the site, NGL plant, product pipelines and terminal facilities shall be carried out to the satisfaction of the Secretary of State prior to the commissioning of the plant. Operation of the facilities shall not begin until any requirements of the Secretary of State in the light of this audit have been complied with."

The Scottish Development Department has said that the Secretary of State will consult HSE and the local authorities before deciding whether the hazard audit has been adequate and the safety level revealed by the audit is acceptable.

This condition is undoubtedly onerous both in terms of the work involved and the time it will take for it to be completed and it is interesting to speculate as to whether some similar requirement might be imposed by planning authorities for other big projects in the future.

Canvey Island. By contrast, the public inquiries at Canvey Island have been partly or even mainly concerned with existing petrochemical installations as opposed to new projects. The story commenced several years ago with an exploratory public inquiry into the desirability of revoking the planning permission given in 1973 to United Refineries Ltd. (URL) to build an oil refinery on Canvey Island. Following a recommendation in the report of that inquiry the Health and Safety Commission was asked in 1976 to carry out an investigation into the risks to health and safety of the various installations there, both existing and proposed.

The report of this investigation was published in 1978 (8) and the public inquiry into the URL proposal was reconvened in 1980 to consider it. In the report on the 1980 inquiry the Inspector gave qualified approval to the URP planning consent but focussed attention on the hazards from the British Gas terminal on Canvey Island.

This terminal was in turn the subject of a fresh inquiry earlier this year, the outcome of which is awaited. The latest inquiry was able to consider an updated report (9) on the installations at Canvey Island which took account of a number of improvements and indeed major changes in plant, equipment and operations of the companies involved.

The protracted inquiries at Canvey Island were notable in that they led to a pioneering exercise in the risk assessment of chemical plants. This provided a potent stimulus to the debate about the practical decisions on the relation between potentially hazardous industry and people who live and work nearby. Again, it has to be said that the considerable benefits which have followed the Canvey investigations have not been bought without substantial cost, both to HSE and the firms concerned.

CONCLUSIONS

Both the benefits and drawbacks of public inquiries into major projects are the subject of much discussion at present. The Royal Town Planning Institute has recently offered some constructive proposals for improvements (7). Michael Mann, QC, a distinguished member of the planning bar, has emphasised (10) that the inbalance that exists between the resources available to the proponents and those available to objectors is probably the most serious immediate cause of concern to the public interest/environmental groups. He considers that it cannot be right in the public interest that the resources available to promote a project should be vastly greater than those available to question its validity. It affects the quality of the evidence that objectors are able to put and must indirectly affect both the Inspector's and the Secretary of State's judgement. There is support for this view in the RTPI and other quarters.

Where safety is an issue at an inquiry and HSE is represented, we see our role as that of a disinterested and impartial adviser. In matters of safety, there is an advantage in having an independent body that one hopes can be regarded as credible and competent, and so to be trusted by those concerned.

REFERENCES

(1) Report: Cmnd 4060, April 1932, Reprint 1966

(2) 206 HL Deb. Col 1

(3) Eg., The Crichel Down affair: Reports: Cmnd 9176, 1954
                                               Cmnd 9220, 1954

(4) Report: Cmnd 218, July 1957

(5) SI 1974, No 419

(6) SI 1974, No 420

(7) "The Public and Planning: means to better participation",
     1982, Royal Town Planning Institute, Portland Place, London
     W1, price £6.00

(8) "Canvey: an investigation of potential hazards from
     operations in the Canvey Island/Thurrock area", 1978, HMSO,
     price £10.00

(9) "Canvey: a second report", 1981, HMSO, price £7.00

(10) "The Big Inquiry: a view from the planning bar", M. Mann QC,
     Symposium on the Big Inquiry, May 1982, Institution of Civil
     Engineers, Great George Street, London SW1.

HAZARDS FROM THE COMBUSTION OF
REFRIGERATED GAS SPILLAGES

J.A. EYRE

PRINCIPAL SCIENTIST
SHELL RESEARCH LIMITED
CHESTER

ABSTRACT

Liquefied natural gas (LNG) and liquefied petroleum gas (LPG) are
expected to become increasingly available over the next decade. As
the utilization of these premium fuels occurs, it is vital that
proper attention is paid to the safety of their storage, handling
and distribution.

Because of the particular properties of the materials, an
accidental spill can lead rapidly to the formation of a potentially
flammable cloud which spreads and disperses under the influence
of gravity and the wind. Experiments have been performed by Shell
Research Ltd., to investigate the dispersion and combustion
behaviour of such clouds. The tests, which were carried out at
Maplin Sands on the north bank of the Thames, involved spilling
quantities of up to $20m^3$ of refrigerated gases onto the sea.
Measurements were made of gas concentration in the vapour clouds
as they dispersed, and in several cases, the clouds were
deliberately ignited so that the intensity of heat radiation and
pressure generated by the flames could also be measured.

The results of the experiments are described and indications are
given of the way in which they, along with results of other parts
of a more widely based safety R & D programme, will be used.

INTRODUCTION

The title of this Annual Conference is "Health and Hazards in a
Changing Oil Scene". The subject of my presentation this
afternoon is concerned with hazards from cryogenic or liquefied
gas spillages and is very much in keeping with the theme of
change; change, that is, from an energy scene which has been
dominated for several decades by oil to one in which other fuels
will play an increasingly important role.

The fuels which I am going to be talking about are liquefied
natural gas (LNG) and liquefied petroleum gas (LPG). Both materials
will become increasingly available over the next decade, and it is
obviously important that as the utilisation of these premium fuels
occurs, proper attention is paid to the safety of their storage,

173

handling and distribution. This is a subject which has raised a
considerable degree of public interest and, in view of the many
dramatic and sensational accident scenarios which have been publicised,
understandable concern.

At this stage, I think it is important that I make clear the
subject of my presentation. Although the titles both of the
Conference and my paper contain the world "hazard", and hazard
has connotations of chance or risk, I will not, in fact, be talking
about the probability of accidents at all. This is undoubtedly a
very interesting subject, particularly the way we perceive and
react to the chance of very low probability events. It is also an
area where I feel more work is required to put hazard analysis on a
sounder technical footing. But today, I will be talking about the
consequences rather than the probability of a spill of refrigerated
gas.

REFRIGERATED GASES

First of all, I should remind you of the properties of the materials
we are dealing with. Both LNG and LPGs consist of mixtures of
hydrocarbons. Methane is the prime constituent of natural gas.
For transportation over long distances, where pipeline transmission
is impractical, or for off-peak storage where an energy-dense
product is preferred, natural gas is liquefied to a product which
occupies only a very small fraction of the original gas volume.
At atmospheric pressure, LNG has a boiling point of about $-160^{\circ}C$.

Unlike methane, the main constituents of petroleum gas, propane
and butane, can be liquefied by pressure at atmospheric temperature.
Alternatively, like natural gas, propane and butane may be
liquefied at low temperature ($-45^{\circ}C$ for propane and $-1^{\circ}C$ for butane)
and at atmospheric pressure. It is this refrigerated material
which I will be concerned with today rather than the more familiar
pressurised form. Both LNG and LPG are carried in bulk by sea from
producing area to market. From the liquefaction plant, LNG is
transported in ships predominantly in the 75,000 to 125,000m^3
capacity range. LPG tends to be carried in refrigerated form, in
smaller vessels, although recent designs have been in the 75,000m^3
range. Land-based storage tanks can have capacities in excess of
100,000m^3.

SPILL BEHAVIOUR

When spilled onto a warm surface, be it land or sea, both LNG
and refrigerated LPG evaporate to form a dense vapour cloud which
spreads and disperses under the influence of gravity and the wind.
A large, potentially flammable cloud can form quickly. Its size
and shape is influenced by several variables including the quantity,
rate and manner of the spill, the topography of the site, the wind
strength and atmospheric conditions, and, in the event of ignition,
the delay between spill and the initiation of combustion. The
consequences of ignition are also influenced by these factors.
Evidence from industrial accidents involving leaks of hydrocarbon

vapour, for example at Flixborough, suggests that combustion may take place at such a high rate that a damaging blast wave is produced which can extend the danger zone beyond the edge of the flammable cloud. This behaviour is rare, but even in the absence of explosion, a large vapour fire could cause extensive damage both within the zone covered by the flame and also at greater distances from thermal radiation.

It is possible to postulate an infinite number of spill scenarios, ranging from a minor leak at, say, a pipe flange, to a catastrophic accident in which, for example, the total contents of a ship's tank of $25,000m^3$ of product was spilled in a very short time. The primary defence against all of these is, of course, in the correct design and operation of refrigerated gas storage and transportation facilitites. Nevertheless, during the planning of a project, information concerning the effect that a postulated spill could have on the environment is required to establish, for example, how far the flammable gas would travel and the consequences of ignition. Together with an assessment of the probability of the events occuring, this information is used to evaluate safety zones around plant, storage facilities, harbour sites and ship routings, etc. It is also used to draw up contingency plans and to design protection systems such as water deluges, catchment bunds, etc.

Thus, in part, the research conducted by Shell companies and others has been aimed at constructing models and correlations that take account of the variables described above and enable us to predict the behaviour of any postulated accident. With this aim in mind, we have recently carried out several experimental programmes concerned with refrigerated gas dispersion and combustion at a scale sufficiently large to provide data adequate to test the models and subsequently to permit any modification and development that might be necessary.

THE MAPLIN SANDS PROJECT

This afternoon, I am going to describe one such programme to you. The experiments, which were carried out at the Proof and Experimental Establishment of the Ministry of Defence at Maplin Sands on the north bank of the Thames estuary, involved spilling quantities of up to $20m^3$ of refrigerated gases (LNG and refrigerated liquid propane) onto the sea. In the majority of the experiments, measurements of gas concentration, temperature, wind and atmospheric conditions were made, as the vapour cloud dispersed over the unobstructed test area. In some of the tests, the clouds were deliberately ignited so that the combustion behaviour could be observed and measurements made of thermal radiation and explosion overpressure.

The extensive tidal mudflats at Maplin provide an ideal site for the experiments (Fig.1). From a shore-based gas handling plant, capable of storing $75m^3$ of LNG or refrigerated LPG, an insulated cryogenic pipeline led to the spill point 350m offshore. Instruments were deployed on 71 floating pontoons deployed in an array around the spill point to give optimum coverage of the likely gas clouds. The signals from the instruments were relayed by cable to computers onshore. In total, there were about 360 instruments in the array including 200 gas sensors, 66 thermocouples, 27 radiometers and 24 pressure transducers. The spills were photographed from three locations: two land-based and the third in a helicopter overhead.

We were able to carry out two types of spill, continuous and instantaneous. In a continuous spill, a steady flow of liquid was established to the sea surface. Evaporation of the liquid produced a gas plume of roughly constant dimensions, connected to the source and existing while the liquid flow was maintained. Instantaneous spills were needed to study the three-dimensional, time-dependent behaviour of heavy gas clouds. In an instantaneous spill, evaporation of the liquid produces a gas cloud which grows with time and becomes detached from the source as it is advected downwind. A submersible barge, which could be raised by pumping air into buoyancy chambers and lowered by flooding the chambers was used for instantaneous spills. As the barge sank, water flooded into it, displacing the liquefied gas onto the sea surface in a period of a few seconds.

GAS DISPERSION

From study of the photographic records of 21 spills, a substantial amount of qualitative information can be derived, for example:
o   continuous spills of both LNG and refrigerated propane gave
     low visible plumes which grew in height downwind, (Fig.2)
o   the plumes became narrower but higher as the wind speed
     increased,
o   in the case of LNG, the plume was more dense and longer when
     the LNG source was a falling liquid stream and shorter and more
     buoyant when the LNG was released under water,
o   under low wind speed conditions, the instantaneous spills gave
     almost circular clouds (Fig.3), which became more elliptical as
     the wind speed increased.
A considerable amount of effort has been, and is still being, applied to assess the quantitative aspects of cloud dispersion. One of the most important tasks has been to compare the measured downwind distance to the lower flammability limit with that predicted by the Shell heavy gas dispersion model HEGADAS. This comparison has shown that, for propane, over a range of wind speeds, there is satisfactory agreement between the experimental results and the model predictions. For LNG, the measured distances are shorter than the predictions, ie. the model is conservative in this case. Unlike propane, methane becomes buoyant as it warms up. The heat transfer from the sea to the cloud seems to be an important factor which must be incorporated into the model and work is currently in hand to achieve this.

It has often been argued that the distance to the time-averaged lower flammability limit (LFL) is not an adequate measurement to use for defining safety distances. Because of fluctuations in concentration brought about by atmospheric turbulence, it has been postulated that a gas cloud may be flammable at distances considerably greater than the predicted, time-averaged LFL. Values of 0.5 LFL or even 0.1 LFL have been suggested as more suitable scales. In one of the continuous propane spills at Maplin, we were able to obtain gas concentration measurements, unaffected by any meandering of the plume, which demonstrated that the fluctuations around the LFL were in the region of 1.4:1 (peak to average concentration). A factor of 1.4 in concentration represents a factor of about 1.2 in dispersion distance.

CLOUD IGNITION

A more direct indication of the distance to the flammable limit was obtained in some of the Maplin experiments. In the case of some established plumes, ignition was achieved somewhere near the centre of the flammable zone and the flame was observed to travel across the plume and both upwind and downwind. The position where the flame extinguished was of particular interest because this could be compared against the gas concentration measurements. For example, in a particular LNG spill of $4.7m^3 min^{-1}$ in a 4.5m s^{-1} wind, the flame was seen to extinguish at 130-140m downwind from the spill point. The distance to LFL based on measurement of peak gas concentrations was 130 $\pm$ 20m in this case. Examination of other plumes gave similar results.

The existence of pockets of flammable gas beyond the edge of the main flammable zone was demonstrated clearly in the Maplin experiments. On several occasions, usually when ignition occurred on the edge of the plume, the flame failed to propagate through the bulk of the vapour cloud. This phenomenon was sometimes seen clearly by observers at the time of the test and /or detected from video recordings. In other cases, such occurrences were recorded by infra-red photography even when flames were not visually apparent. It would seem that while pockets of gas may exist beyond the time-average LFL as described above, from a hazard assessment point of view, this extension of the flammable zone should be coupled with a reducing probability that the entire cloud will burn. Intuitively, it would seem that the chances of ignition occurring in the first place must also decrease in this ill-defined region at the edge of the cloud.

COMBUSTION BEHAVIOUR

Eleven combustion tests were carried out at Maplin comprising both instantaneous and continuous spills of LNG and refrigerated liquid propane. As with the dispersion tests, a considerable amount of qualitative information can be gleaned from the photographic records of the tests:

o   Approximately half of the successful ignitions failed to propagate through the bulk of the cloud.

o   Ignition of LNG clouds took place within the visible fog; the
downwind visible edge extended beyond the position where the gas
had diluted below the flammable concentration. The converse
was true for propane, ie. the visible edge lay within the
flammable region.

o   Immediately following ignition, combustion of the leaner
pre-mixed portion of the cloud was characterised by a relatively
fast, weakly-luminous flame. (Fig.4)

o   In the continuous spills, the richer parts of the plume burned
with a bright yellow wall of flame which propagated towards the
spill point. The rate of propagation was low and indeed a wind
speed of 4-5m s^{-1} was sufficient to hold the flame virtually
stationary in the case of LNG.

o   The flames increased markedly in height as soon as the fire
burned back to the liquid pool at the spill point. Flames up to
100m high were seen. Refrigerated propane pool fires produced
very sooty flames (Fig.5), whereas an LNG pool fire was soot free.

Measurements were made at Maplin of both flame-generated overpressure
and thermal radiation, these being the two primary effects which
can extend the danger zone beyond the edge of the flammable cloud.

Theoretically, damaging overpressures would be produced if the cloud
deflagrated with a flame speed in the order of 150m s^{-1}. At Maplin,
the fastest flame speeds were observed during the combustion of the
pre-mixed portion of propane clouds. Typically, these speeds
averaged 12m s^{-1}. Higher transient speeds up to 28m s^{-1} occurred,
but no sustained accelerations which could have led to appreciably
faster flames were detected. Similar behaviour was observed for
natural gas clouds but the flame speeds were lower.

Overpressure generated by the flames was measured in several tests
but always at a very low level. The maximum overpressure detected
was approximately 1 mbar which is well below the damage threshold
for either structures or people.

It must be stressed that higher flame speeds and pressure may result
from confinement, turbulence produced by obstacles in the flow field
or from more energetic ignition. These factors were not included in
the Maplin tests but further work to investigate them is planned.

In the absence of damaging overpressures, the primary hazard from a
refrigerated gas spill is that of fire. The danger could extend
beyond the boundary of the flammable cloud, both because of
expansion of the hot products and because of thermal radiation to
the surroundings. At Maplin we saw that the hot products expanded
vertically. There was little horizontal expansion. This is
important because it means that the primary fire zone (excluding
radiation effects) is defined by the pre-combustion LFL contour,
subject of course to the concentration fluctuations discussed
above.

Radiation measurements were made at Maplin by means of wide-angle radiometers which were deployed so that they viewed the whole of the burn. The radiation received at a target depends on several factors including the distance between the fire and the target, the orientation of one relative to the other, the humidity of the atmosphere between the two, the duration of the fire and its shape, size and emissive power. By assuming that the fuel-rich fires behaved as surface emitters, it has been possible to calculate the surface emissive powers of both cloud fires and pool fires from the Maplin data. LNG and propane cloud fires gave similar values of surface emissive power. The single LNG pool fire gave a slightly higher value and the refrigerated propane pool fire gave a considerably lower value because of the dense clouds of soot which masked the flame.

SIGNIFICANCE OF THE MAPLIN TESTS

A considerable volume of information has been accumulated from the Maplin experiments. In general terms, it can be seen that the results permit a more detailed understanding of refrigerated gas spills under relatively simple and well-defined conditions. It is, of course, inevitable that there will be differences between idealised experimental conditions and situations that may arise in practice, but until an adequate understanding of the simple case has been achieved, it is much more difficult to predict behaviour under more complex circumstances.

Thus, the Maplin tests have provided a comprehensive set of dispersion data which is being used to assess models, thereby greatly improving our confidence in their applicability to larger spills and the accuracy of calculated safety distances. The same is true of the combustion experiments, which, in addition, have provided information that is being used to improve the safety of plant and ships to fire and explosion. Further valuable, practical information has been obtained relevant to contingency planning, for example, to cargo jettisoning and fire-fighting procedures.

Analysis and application of the Maplin results are still underway. Inevitably questions remain to be answered, but we believe that such work can go a long way to dispel much of the concern surrounding the development of clean, safe and practical alternatives to oil.

ACKNOWLEDGEMENTS

The author wishes to thank Shell Research Ltd. for permission to present this paper and various colleagues for helpful advice during its preparation.

The author is also pleased to acknowledge the assistance of numerous individuals and organisations, both inside and outside Shell companies, in planning and carrying out the work described in this paper.

180

(Further information is available in Paper 29 entitled 'Dispersion
and Combustion Behaviour of Gas Clouds resulting from Large
Spillages of LNG and LPG on to the Sea' by D.R. Blackmore, J.A. Eyre
and G.G. Summers which was given at a joint meeting of the Institute
of Marine Engineers and the Royal Institution of Naval Architects
on 10th May 1982.)

Figure 1 - View of the Maplin site showing base camp, gas storage
tanks, spill pipe and pontoon array.

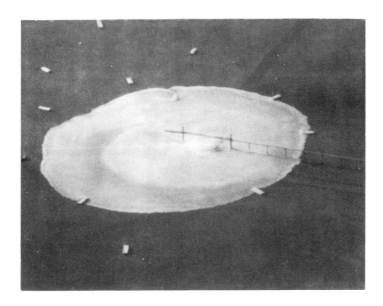

Figure 2 - The plume formed during a continuous spill of LNG in a
moderate wind; spillage 2.5m^3/min.

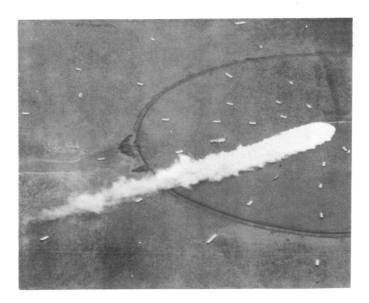

Figure 3 - The circular cloud formed during an instantaneous spill of
refrigerated liquid propane under low wind speed
conditions; spill volume 27m^3.

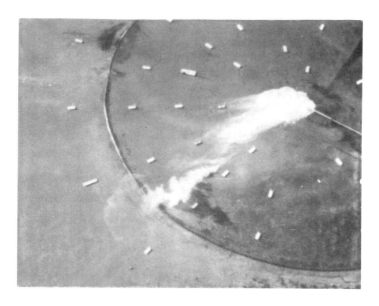

Figure 4 — Flame development during the combustion of the gas plume from a continuous spill of refrigerated liquid propane; spill rate 5.6m^3/min.

Figure 5 — The pool fire formed when the flame has propagated back to the spill point during a continuous spill of refrigerated liquid propane; spill rate 5.6m^3/min.

B.W. Eddershaw (Safety Group, ICI Petrochemicals and Plastics Division

I congratulate Dr. Eyre on his presentation, which was full of information . Could he say whether the experiments confirm that which some believe, namely that unconfined clouds are difficult to ignite (and therefore will not easily explode)?

He did say that the confinement needs to be high in order to accelerate a flame front rapidly enough to cause an explosion. Thinking about the application of the results of the experiments to onshore conditions, what degree of confinement would he think would be necessary?

Dr. J.A. Eyre

In answer to the first part of the question, I think it would be misleading to suggest that unconfined vapour clouds are difficult to ignite. It is true that a certain combination of circumstances, ie. gas concentration, flow rate, ignition energy, etc. need to be satisfied before ignition will occur, but such condition can be readily achieved in practice. The observation at Maplin, that on several occasions a portion of the cloud was ignited yet the bulk of the cloud failed to burn, is probably an indication of inhomogeneity in the gas concentration round the lower flammable boundary. Another possible contributory factor is that the strong buoyancy produced in the ignition region might, under some circumstances, be sufficiently strong to detach the flame from the cloud.

The degree of confinement necessary to bring about explosion has been investigated in several small-scale studies. In general, it has been found that a considerable degree of confinement, for example, an open-ended pipe or largely enclosed box, is required to produce pressures capable of causing structural damage. Whilst there is some evidence that the severity of the explosion (ie. the rate of combustion) increases with scale, the presence of obstacles and other flow-disturbing mechanisms is probably more important but, as yet, unquantified.

MATERIALS HANDLING AND ECONOMICS

P.R. Davis
Professor of Human Biology and
Director, Materials Handling Research Unit,
Robens Institute, University of Surrey.

INTRODUCTION

This paper reports some of the preliminary results of a survey of
materials handling and ergonomics in the British oil industry, the
main work being in the charge of my collegue, Mr. Geoffrey David,
of the Robens Institute, who has been given a three year Shell
fellowship to this end. The objective of the work is to obtain
systematic analyses of the requirements, and to identify areas of
danger or difficulty, in the handling of materials from their
original production to their final use by the consumer.

The Institute staff have very large experience of industrial
handling problems, having carried out similar surveys for the
Ministry of Defence, the Electricity Council, the Post Office,
and many other large and small concerns, so that the basis of the
methodology for such a large scale activity is well founded. I
must emphasise that it is the aim of the fellowship to identify
problems, not necessarily to provide answers at this stage,
although answers to problems will be sought from appropriate
agencies.

The methods we have developed for these purposes lie in three
overlapping areas, epidemiology, physiology and ergonomics.

EPIDEMIOLOGICAL DATA

For epidemiology we now have a data bank concerning some 2 million
workers in a wide variety of occupations which we use for comparative purposes. To use data in this way, we find it necessary to
compile profiles of each occupation, which tells us the total
number employed in each grade, their age structure, their length
of service and, if possible, their training record, together with
details of their employment, such as the shift pattern, and weekly,
monthly and annual trends, as well as the natures, times, severity,
and consequential time losses of illnesses and injuries.

In most industries, these details are available, but are commonly
held by different departments. Non-time lost injury data are often
held by the factory nurse, who may also hold details of absences
under 3 days. Three day plus data is often held centrally, but the

185

details of each incident may again be present only in the aid post. The numbers employed, and details of employees experience, and length of service, are commonly held by the staff office or in accounts, who may have no way of cross-reference with the medical service records. Some departments may be computerised and others not, and the computers in one part of an organisation may not be able to talk to those in another part. Thus to obtain a proper profile of the work force and their frailties can be sheer grind, but is usually a most enlightening exercises. I hope two examples will show the value of such data. Firstly, if one takes the age structure of telecommunications engineers, and compares this with the age structure of those with 3 day plus injuries, one finds that there is a significantly high rate in the middle age group. If one does the same thing with construction engineers, it is clearly the young that are at risk. The basic reason for this difference is that the telecommunication engineers undergo rigorous training, are properly supervised in their early years of practice, and the more difficult and dangerous tasks are undertaken only by experienced men, whereas the construction workers have little training, and the older man gives the youngster the dangerous tasks. In telecoms the loss of trainee workers is low, and the overall injury rate is kept down in most trades. In construction the loss of trainee workers is high, and the overall injury rate is very high throughout, as is the loss of experienced workers from the industry.

A second example is a comparison of injury rates of those in different occupations. Provided the data are soundly based, and the job classification is realistic (and titles such as "engineer" can mean anything and have to be amplified), one obtains further valuable information, and can immediately suggest those areas of work most likely to repay further detailed study.

When analysing data concerning the nature of injuries, we find it of value to think of them under two headings, these being:

Non work-connected. (Minor infections, dental troubles, social diseases, etc.)

Possibly work connected. Acute disorders (cuts, bruises, burns, fractures, etc.) caused by immediate hazards.

Cumulative disorders (backs, "tennis elbow", tenosynovitis, bursitis, certain eye conditions, etc.) caused by repeated over-stress over considerable periods of time.

In any work situation, one finds a basic rate of acute disorders, however safe one tries to make the job, simply because workers are human beings and we all do silly things. One can expect some 30/1000/year of acute disorders, and some 2-5/1000/year of cumulative disorders to occur in the most harmless occupations. It is when one finds increases above these sorts of base levels that one needs to consider causation in detail. In this regard, for materials handling studies, we have found the base rate to have general value, as in most occupations so far investigated, a significantly high base rate is always accompanied by significant increases in acute injuries due to materials handling, and usually indicates that considerable reduction in morbidity can be achieved at little cost, with consequent increase in productivity.

By analysing the sequence of events leading up to the accidents, one can sometimes highlight basic causes apart from those stated in accident forms. For instance, where there are a lot of slips and mishandling occurences in a given location, one suspects the footing, the lighting or the rate of presentation of work, and linear analysis can often differentiate between these.

ACCIDENTS AND INJURIES

Having identified areas of concern, one can then start to observe the activities in those particular occupations, and to seek the reasons for the unwanted injuries and absences. For acute injuries this can often be a fairly simple task, but the causes of cumulat- ive tensions can be more difficult to identify. We find it necessary to use a number of physiological techniques, and we have had to develop methods which require as little interference as possible with the workers activity. Since cumulative injuries result from repeated overstress, one has to measure the forces that require to be exerted, and one has to have knowledge of the force levels that can be applied safely by the work force in question. To this end, we have used an indirect method of measure- ment of internal reaction to physical stress, which results from the well established observation that there is a rise in pressure in the trunk cavities which accompanies any forceful exertion, and that this rise in pressure is linearly related to the magnitude of the forces acting on the lower spine. Any force exerted by the hands has to be matched by equal or opposite forces at the feet when standing, or the buttocks when sitting, and the trunk is the necessary link between the hands and these other points. Thus measurement of the forces acting on the trunk gives a reliable measure of the forces required in virtually all manual tasks. Further large scale studies of the forces applied in jobs with and without high accident rates have shown that, for males, those occupations requiring persistently repeated trunk pressures about 90 mm/Hg have a significantly high incidence of back disorders and accidents.

Using this relationship, it has been possible to create diagrams of forces* which can be applied by male workers without undue risk, and one can then use these to establish the acceptability of existing work situations. The present diagrams cover pulling, pushing, lifting, and rotating thrusts, using one or both hands, when standing, squatting, sitting or kneeling, so that the very large majority of tasks can be assessed without any disturbance

*Force Limits in Manual Work, M.H.R.U., University of Surrey, I.P.C., 1980.

of the worker, and have been adopted for use in the European Coal
and Steel Community. Where there is the occasional complex task,
then it may be necessary to carry out direct measurement, in which
case we use a telemetering system which again causes little interference.

In general, where a task is accompanied by a high accident rate, one
usually finds a number of elements requiring high stress force
applications, and one can identify those instants within the task
inducing stresses above the critical level. It is then a matter of
correcting the situations as far as possible, either by altering
the working method, or some environmental change, or by more radical
design changes.

ERGONOMICS

So far we have considered existing work situations. The ergonomic
approach is properly more concerned with designing for future efficiency and safety. It requires a proper systems approach, and can
give guidance to the designers of equipment of work stations. It
is concerned not only with the task in hand, but of the wider needs,
such as the requirements for efficiency and safety right from the
start to the use of the final product by the consumer. Thus it has
relevance not only to the factory that may be in question, but also
to the producers of raw materials and to the design of the consumer
product and its use in the community.

Thus, ergonomics is a matter of planning for human inputs at the
start of the design phase. While it can contribute to solutions to
problems in existing plant, when a factory or work station has been
provided it is usually too late to take full advantage of ergonomic
knowledge, and thus our present fellowship can only seek to identify
major areas of concern within the existing frameworks.

THE OIL INDUSTRY PROJECT

Results So Far

It must be stressed that at this stage in our project our results
are generally not far advanced, and our knowledge in some areas is
better than in others. Thus the state of our analysis is patchy,
and in some cases may not be fully representative of the industry
as a whole. However, we do have some significant observations
already, and I will briefly present these here, stressing that they
are far from comprehensive at this stage.

Offshore Exploration And Supplies

The average lost time injury rate is about 100/1000 men/year, which
is three to four times greater than the basic rate in non-hazardous
industries. The injury rate seems to be much higher for contract
labour than for those directly employed by the oil companies, the
ratio being 14/1 for offshore operations, and 3.5/1 in shore stations. There is clearly a perturbation of reporting of minor injuries
in offshore installations, in that the ratio of lost time to non
lost time injuries is about 1/45 in most onshore activities, but is
only 1/1.5 on the rigs. Serious injuries on the rigs vary between
occupations, being:

/1000/yr.

Drilling	167
Maintenance	165
Construction	102

and marine supply has a relatively high rate of 72/1000/yr.

As we all know, falls offshore are relatively common, accounting for
some 40% of lost-time accidents, as compared with some 25% in other
occupations, although the number of accidents due to stepping on or
striking against objects is relatively lower at 12%. 30% of offshore accidents are associated with equipment handling, as opposed
to equipment failure, and reflect the local storage and circulation
limitations.

On the supply side, tripping accidents are common, and not surpris-
ingly, falls on board supply ships are relatively frequent.

The high rate in maintenance workers reflects that found in many
other industries. With the emergence of greater understanding of
human industrial capacities, the design of much production machinery in industry generally has been greatly improved of recent years,
resulting in significant decreases in injury rates, and improvements
in production performance. But very little thought has yet been
given to design for maintenance, and with the coming of automation,
the proportion of maintenance workers is increasing rapidly, and
almost everywhere it is maintenance workers who have the highest
casualty rates.

Those figures we have so far obtained for marine supply show a
significantly high rate as compared with other marine organisations,
although the onshore supply figures are much better. One of the
reasons for this is the rapidity with which the North Sea had to be
developed. This led inevitably to the grafting of high intensity
supply organisations onto declining fishing ports, resulting in
much Topsy-like growth and difficult road and rail communications.

We have seen men wearing hard footwear having regularly to stand
on, wobble on, and risk damage to expensive valve parts to reach
small items of equipment called for frequently by the rigs: we have
seen containers being loaded onto ferries nearly full of vegetables,

but with large unprotected drilling bits loaded on top of the sacks
Had vinegar been added, a large supply of cole slaw would have been
available by the time it reached the rig. We have seen parts stores
in which the most frequently required items were furthest from the
supply counter. We could find no analysis at local level of freq-
uencies of use, magnitudes of load, or storage needs for very many
kinds of equipment. Thus, while in the exploration and production
area, our studies are only at a very preliminary stage, we are confident that there is scope for more in-depth investigations of various aspects of materials-handling, for example in warehousing
strategy, storage needs and transfer arrangements.

In refineries. Serious incidents are uncommon, being as low as
6/1000/yr. for oil company personnel, and 22/1000/yr. for contract
labour. However, minor injuries are relatively frequent, 165/1000/
yr. for oil company personnel and 542/1000/yr. for contractors'
staff. Of those occupations so far studied, gantry workers and
maintenance engineers have the highest injury rates. One feature
of concern is maintenance levels for valves.

Distribution. On the distribution side, drivers, vehicle mechanics
and terminal operators are clearly at some risk, with a lost time
rate of between 30 and 70/1000/yr. and a significantly high back
rate. Falls and trips account for over half these injuries, and
handling is associated with a further 20%.

We have carried out some fairly detailed studies of distribution
operations, and it is clear (as you all probably know already) that
the road tanker cab egress and ingress will repay study, as will
the design and location of outlet faucets. A considerable proportion of incidents occurs in association with access to work on tank
tops. Further, the design of discharge points, and manholes and
covers are susceptible to considerable improvement.

Studies of delivery drivers at work have shown that they carry out
some 400 stooped force applications per 8 hour shift, some 200 of
which could be avoided by alterations in design. Many of these
lifts require heavy force applications. There are roughly 1.2
million deliveries per year of retail motor spirit alone, and we
are therefore talking about 240 million heavy force applications
which could be avoided by systems design.

You will ask what is the purpose of all this effort? Can it be
effective? The answer is yes it can be. We have applied our
methods to an oil processing subsidiary, whose role was to receive
refined products, blend them and package them for the retail market.
They were in an industrial complex in which, unknown to us at the
time, a rival university was carrying out a careful longitudinal
survey of industrial accidents and illnesses. As a result of our
studies, we made recommendations for modest alterations in their
working methods, most of them requiring no capital expenditure.
The factory adopted all our suggestions, and their survey team
recorded a 70% drop in injury rate and absence, which has been

maintained for the last 4 years.

As you see, our results are patchy, and although we have more data than that reported here, it is not yet digested adequately for presentation. However, I welcome the opportunity given by the invitation to speak here to present our raison d'être and our methodology. I also must express our deep gratitude to Shell UK Ltd. for their support, while at the same time explaining that our observations and findings are not just Shell based, but extend to the industry as a whole. Finally, I hope that our efforts will in some way contribute to future safety and efficiency in the industry. The British oil industry has responded most vigorously to the call for North Sea development, and has had to rush through immense development programmes. At this moment in time, it is concerned to consolidate its gains and to plan for the future. In planning for the future, I suggest that the lessons of the past as shown by our studies of existing installations could be of value, and that the ergonomics appraisal which is just beginning will be of very considerable benefit, both in terms of production economy and of human safety.

HAZARD SURVEY AND CONTROL

R.H. SMITH

ESSO PETROLEUM COMPANY LIMITED

INTRODUCTION

The purpose of this paper is to identify, in the area of Major
Hazards, what has already been done by Industry and the
Legislators, and the issues that have to be tackled in the near
future. In this paper the widely used term 'Major Hazard' is
considered to refer to those industrial installations which handle
hazardous substances and have the potential to put the public at
significant risk as a result of a serious incident, such as an
uncontrolled release of flammable or toxic material. The paper does
not cover the environmental and employee occupational health aspects
of these activities.

The subject will be considered in five sections:-

1. The background to the development of proposed regulations
   together with an explanation of terminology.

2. The changes that are taking place in Industry and the Public
   Sector.

3. Control of Major Hazards in the Petroleum Industry.

4. A brief discussion of Hazard Assessment techniques.

5. Identification of the issues that we have to manage in the near
   future.

1. Background to the Development of Regulations

In the U.K. Lord Robens chaired a committee which submitted in 1972
a Report "Safety and Health at Work". As a consequence, the Health
and Safety at Work etc. Act was introduced in 1974 and placed
significant responsibilities upon employers "to ensure so far as is
reasonably practicable, the Health, Safety and Welfare at work of
all his employees" and "to ensure, so far as is reasonably
practicable, that persons not in his employment are not exposed to
risks to their health and safety". Duties are also placed on
employees and on manufacturers. Throughout the Act there is an
emphasis on actions that are reasonably practicable. The
Flixborough incident in 1974 led to establishment of the Advisory

Committee on Major Hazards (ACMH) by the Health and Safety
Commission. The Committee was asked to consider the safety
problems associated with large scale industrial premises conducting
potentially hazardous operations. The Committee published reports
in 1976 and 1979 which recommended, and provided the basis for,
practical regulations for Notification and Survey of Installations
carrying out potentially hazardous activities. The concepts of
Notification and Survey are detailed later in the paper.

Regulations were then developed by the Hazardous Installations
Policy Branch (HIPB) of the Health and Safety Executive (HSE) and
published as a Consultative Document in 1978, and were essentially
agreed by late 1980. UK Industry recognises the need for control
regulations, and has been involved in considerable consultation
and support through the Industry Associations, the Institute of
Petroleum and the Confederation of British Industry.

In Europe, following the incident at Seveso, Italy in 1976, the
European Commission published in 1979 a draft Directive concerning
the Major Accident Hazards of Certain Industrial Activities. As a
results of experience in the U.K., the HSE were able to effectively
influence its development. The Directive has suffered considerable
delay, but was approved in late 1981 and adopted in June by the
European Council of Ministers. The sequence of these developments
is illustrated in Table 1.

In order to progress recommended changes to Planning Procedures in
the U.K., the two parts of the proposed Notification and Survey
Regulations have been segregated to enable rapid progression of the
Notification Regulations, and to ensure that the subsequent Survey
regulations will be no more onerous than the requirements of the
EEC Directive.

PROPOSED REGULATIONS FOR NOTIFICATION

In the U.K., Notification Regulations will take effect on 1st.
January 1983, and Notification will be required by 31st. March.
Notification will require submission of data on certain inventories
which are liable to be held on a site. This is essentially an
identification process which will alert the HSE to the existence of
sites with notificable inventories. It will then enable the HSE
to given appropriate attention to activities and developments on the
site, and also enable them to give appropriate advice to Local
Planning Authorities on activities and developments in the vicinity
of the site.

Examples of the substances that will be notifiable, and the inventories
(in storage and processes) at which they become notifiable are:-

Flammable Gases	15 tonnes
Pressurised LPG	25 tonnes
Refrigerated LPG	50 tonnes

Hydrocarbons (flash point below 21°C)	10,000 tonnes
Hydrogen Fluoride	10 tonnes

It is foreseen that, when UK Regulations to implement the EEC Directive are in place, formal Survey reports will need to be submitted to HSE in the period 1984-89. Survey would be required at inventory levels ten times those requiring notification as tabulated above, and for any other substances not normally handled in the Petroleum Industry. Hence refineries and the larger terminals and storage sites would become involved. The content is not yet finally established but basically each site management would be required to identify the nature of the potential incidents and their possible consequences, and to demonstrate that proper systems are in place to prevent or control such incidents. The aim is to establish a reasonable and generally improved method of site and emergency control.

2. THE CHANGES THAT ARE TAKING PLACE

The HSE has established the Major Hazards Assessment Unit (MHAU) in order to monitor or carry out hazard surveys or more detailed risk assessments for existing sites or new project proposals (for example Canvey Island and Moss Morran), to give guidance to HSE field inspectors, and provide advice to Local Planning Authorities to assist in their development of local planning strategy. A fundamental long term national objective, which must be supported as being prudent, is to avoid placing more people at risk, and preferably to reduce the number at risk. A number of assessments have now been carried out in the UK in varying detail. This work has resulted in a much increased public and political awareness of existing or proposed installations which carry out an activity with potential to create a major hazard.

The downstream petroleum industry itself is now undergoing significant change as a result of a substantially reduced and changed demand pattern. Competition is fierce, and companies are having to rationalise their refining and distribution facilities, while investing in conversion processes to upgrade fuel oil to transportation fuels and special products (eg. lubricants). These changes will involve Industry in a large amount of work, so industry does expect realistic and practical regulations that will be implemented in a reasonable and flexible manner.

3. CONTROL OF MAJOR HAZARDS IN THE PETROLEUM INDUSTRY

The policy of the Petroleum Industry is to protect its employees, the public, and its capital equipment by using the best practicable and cost effective methods and to meet legislative requirements as a minimum. The companies have basic beliefs that most incidents and injuries need not occur, and that self regulation is an effective means of implementing these policies.

The downstream petroleum industry has around 100 years of international experience, which has led to development of well defined practices which cover:

- Equipment design and spacing
- Procurement and construction of equipment
- Commissioning, operation, inspection and maintenance
- Incident Control facilities (eg. isolation, depressurisation)
- Emergency procedures, both onsite and involving offsite authorities
- Management audits and safety surveys
- Incident investigation and consequent recommendations for action

Our experience leads to development of improved procedures and training, and update of design practices. Installation of new plant, or modification of existing facilities, therefore achieves a progressive reduction of risk.

Individual company practices are supplemented by Industry Codes of Practice, such as those published by the Institute of Petroleum and LPG Industry Technical Association.

All companies use a combination of techniques to control risks, but the evolution and structure of the different companies does vary, so that there are some differences between the practices used to achieve the same objectives. In general, however, a pragmatic approach, to identify potential incident sources and take rapid action to establish control, has been found to be very effective. The results is that the record of the Petroleum Industry is extremely good. We believe that the objective of the ACMH would be achieved if all industries reach the best standards of hazard survey and control that are currently practised in the petroleum industry. This would represent a marked general improvement in the U.K.

Our experience shows that the more probable, or credible, event such as release from small bore pipework, create a relatively small consequence, since response systems are in place to prevent escalation (the so-called domino effect). It is believed that the major incidents such as large explosions are extremely unlikely to occur, as a result of continually improving inspection procedures, isolation equipment and emergency facilities. It is extremely unusual for the public to be seriously affected by a major incident within an operating site. However, industry recognises the possibility, and has developed and tested emergency plans with the local fire, police and medical authorities. The total Major Hazards activity is aimed at making these extremely unlikely incidents even more unlikely.

4. THE OUTLOOK FOR HAZARD SURVEY AND ASSESSMENT

Within our industry, risk assessment is mainly applied to new
facilities or major developments in refineries, major distribution
terminals, pipelines and LPG facilities (bulk and customer storage).
However, use of the technique by HSE is moving to existing facilities,
and in the near future, may possibly extend to transportation by
road, rail and sea.

It is foreseen that UK survey regulations will require hazard surveys
of major installations within the next 7 years, and indeed the
Petroleum Industry has indicated to HSE that priority surveys can be
undertaken early in this period. It is expected that these surveys
will have to be updated periodically, and furthermore that any of
these may give rise to a more detailed assessment on specific
aspects of the survey. This would be in terms of probability of
occurrence and consequences, using appropriate and relevant
quantitative techniques.

Understandably, many people would like to see the risk quantified in
all cases, but detailed Risk Assessments are extremely time
consuming and hence expensive (circa $£\frac{1}{2}$ million). Industry has yet
to be convinced of their value in any other than limited and well
defined circumstances for very specific purposes. Little incident
data is actually available to provide adequate statistics and hence
data and model assumptions do need to be made. Much of the
technology is not yet rigorously established and hence calculation
of consequence should not be assumed to be well founded. Industry
cannot support elaborate calculation based on incomplete data and
unsatisfactory assumptions. This is being increasingly recognised,
and we hope realism will prevail with regard to just what can be
calculated, and its inherent accuracy.

5. THE ISSUES THAT HAVE TO BE MANAGED IN THE NEAR FUTURE

Firstly, regarding the content of Hazard Survey, Industry and HSE
need to agree what is necessary and meaningful to ensure necessary
actions or precautions can be identified. We also need to establish
priorities for survey of existing or proposed installations, in order
to effectively use our resources in terms of both cost and skilled
manpower, especially in the current period of substantial change. We
should not divert resource from safety management and improvement
into unreasonably detailed risk analysis.

Secondly, industry and HSE will soon be working together to identify
how to use and publicise the results of these surveys and
assessments so that their meaning can be understood without causing
unnecessary alarm. The proposed European Directive will require
release of information to the local authorities and public, and
increasingly hazard assessments will be an essential input to local
government procedures to establish short and long range land use
strategy. Industry strongly believes that consultation with the
site management is essential, in order to take local conditions and
relationships into account, and to put the risk into context, by

identifying the control methods in place and its low probability.
Consultation will provide an opportunity to consider possible
changes to any planning proposals.

Thirdly, we believe it is important to avoid excessive delay or
cost to projects, by ensuring that only the essential elements are
comprehensively assessed, that the time span of enquiry is limited
without being undemocratically restrictive, and that only cost
effective additions are progressed.

Fourthly, industry needs to be able to develop new and existing
sites without unreasonable constraint. Hence the oil industry must
maintain a high performance standard and demonstrate use of best
current, but realistic, practices. Also local authorities will
need a long term planning strategy to avoid incompatible developments,
while retaining a flexibility to balance the local industrial and
economic interests.

Finally, we will need to respect the competitive position of the UK
by encouraging consistent implementation of the European Directive
by all Member States in order to achieve European harmonisation.

SUMMARY

There is an increasing public and political concern regarding
potentially hazardous installations. Industry will continue to
support the development and effective implementation of practicable
regulations and planning procedures through consultation. At the
same time it needs flexibility to adapt to significant market
changes, and to ensure the efficient use of its skilled resource.
Hence industry wishes to ensure that hazard surveys will be
objective and cost effective. Industry has many control mechanisms
in place, but will need to continue to demonstrate a high standard
of performance and use of best realistic practices. Local
authorities will make increasing use of assessments to establish
long term planning strategy. Publication of the results will need
to be carefully handled to place the risk into context, avoid
unnecessary alarm, and enable planning authorities to balance
various local interests.

TABLE 1. Development of regulations

U.K.   1972   Robens Report

          1974   Health and Safety at Work Act - emphasis on
                    what is reasonably practicable

          1974   Flixborough Incident

          1974   Advisory Committee on Major Hazards reported in
                    1976 and 1979

          1978-80   Consultation on HSE Regulations - considerable
                    industry support

Europe   1976   Seveso Incident

          1979-82   Development of European Directive

Supporting Discussion to Mr. Smith's Paper:

P. Jones (Institute of Petroleum)

Mr. Smith stressed the quite notable safety record of the petroleum industry, due to its stringent codes of safe practice - a very major and ongoing activity of the Institute. These facts are borne out in the paper that those of us who were here yesterday will have heard delivered by Dr. Alderson. While primarily an epidemiological survey of health within the petroleum industry in the UK, his tabulation showed that for the 25 year study: from 1950 to 1975 of eight refineries (comprising a bulk of UK refining capacity in this period), there were 12 deaths by accidental fire and explosion - one accidentally outside of work. For the petroleum bulk installation study comprising some 700 distribution centres over the same period of 25 years Dr. Alderson and Dr. Rushton in their report record total deaths of 7 from this cause (inclusive of drivers of road tankers conveying petroleum product of all categories attached to these centres); of these 7 deaths one was in a domestic fire at home and one occurred in an accident at a subsequent place of work.

Sir Douglas in his opening address spoke of the variability in approach to the matter of what is the cost basis for avoidance of a potential death: I would support Mr. Smith in his qualification that risk analysis must be kept in prospective, as I would doubt whether any of the above eleven deaths could necessarily have been 'designed out' of the systems concerned.